CONTENTS

About the Author 5

Glossary & Abbreviations 6

Maps 8

Bases 11

Preface and Acknowledgments 12

Introduction 14

Chapter 1 April to July 1941 • Humble Beginnings 17

Chapter 2 August to December 1941 • Sweeping the Skies 31

Chapter 3 January to June 1942 • An Increasing Pace 49

Chapter 4 July to December 1942 • A Change of Scenery 61

Chapter 5 January to June 1943 • Facing the Japanese 69

Chapter 6 July to December 1943 • The Days are Not So Easy 87

Chapter 7 January to December 1944 • Into the Fray Again 97

Chapter 8 January to December 1945 • To the War's End 109

Appendix 1 No. 452 Squadron Commanding Officers 116

Appendix 2 No 452 Squadron Honours & Awards 117

Appendix 3 No. 452 Squadron Aces & Claims 118

Appendix 4 No. 452 Squadron Tally Board & Losses 120

Appendix 5 No. 452 Squadron Roll of Honour 128

Sources 132

Index of Names 134

The badge of No. 452 Squadron. The Latin motto is Mundus Pro Nobis, which means "Sweeping the World Before Us".

No. 452 Squadron Battle Honours

English Channel and North Sea 1941-1942

Defence of Britain 1941-1942

Fortress Europe 1941-1942

Darwin 1943 -1944

Pacific 1943 -1945

Morotai 1945

Borneo 1945

Dedication

To the memory of:

Squadron Leader Keith "Bluey" Truscott, DFC*, MiD, RAAF

Wing Commander Brendan "Paddy" Finucane DSO, DFC**, MiD, RAF

Squadron Leader Ray "Throttle" Thorold-Smith, DFC, RAAF

All of whom did not survive the war, and also to the memory of the other 47 members of No. 452 Squadron that also did not return.

SWEEPING THE SKIES

A History of No. 452 Squadron RAAF 1941–1945

MARK LAX

Sweeping the Skies
A History of No. 452 Squadron RAAF 1941-1945
Mark Lax

ISBN: 9781764193771

First published 2026 by Avonmore Books
Avonmore Books
PO Box 217
Kent Town
South Australia 5071
Australia

Phone: (61 8) 8431 9780
avonmorebooks.com.au

NATIONAL LIBRARY OF AUSTRALIA

A catalogue record for this book is available from the National Library of Australia

MIX
Supporting responsible forestry
FSC® C124385
FSC www.fsc.org

Formal approval to use the No. 452 Squadron badge in this publication has been received from the Air Force Brand Manager.

Cover design & layout by Diane Bricknell

Aircraft profile artwork by Juanita Franzi

Front Cover: the artwork depicts a No. 452 Squadron Spitfire over the skies of Darwin in 1943. It was painted by the late Peter Connor, an artist employed by the Department of Defence, and featured in an RAAF News calendar in the 1980s.

Back cover: A58-354 showing off the clean lines of the Spitfire Mk VIII and the tropical air filter fitted to the nose.

ABOUT THE AUTHOR

Dr Mark Lax is a retired air force officer who joined the RAAF in 1974 and retired in 2024. He had a wide and varied flying career including test flying the F-111. Later he had senior staff positions in Air Force HQ and in HQADF. Subsequently, he reinvented himself as a military historian, educator and author.

In later years, Mark has continued to provide specialist advice to the Office of Air Force History and is one of the authors of the official history of the ADF in Afghanistan and Iraq where he wrote the air force chapters. He now writes and publishes widely.

He is a graduate of the Universities of Melbourne and NSW and of the Kennedy School of Government, Harvard University.

Mark is married to Margaret, and they have two daughters and three grandchildren.

Both Mark and Margaret are Legatees with Canberra Legacy and Mark is currently Chair of Legacy Australia.

Three key leaders of No. 452 Squadron in the UK circa 1941. Left to right: Squadron Leader Keith "Bluey" Truscott, Wing Commander Brendan "Paddy" Finucane and Squadron Leader Ray "Throttle" Thorold-Smith.

GLOSSARY & ABBREVIATIONS

ARD	Aircraft Repair Depot
ATC	Air Traffic Control
Babs	Allied codename for Mitsubishi C5M reconnaissance plane
Betty	Allied codename for Mitsubishi G4M bomber
Circus	RAF codename for an aerial sweep over enemy territory designed to draw up enemy fighters.
CO	Commanding Officer
Dinah	Allied codename for Mitsubishi Ki-46 reconnaissance plane
DFC	Distinguished Flying Cross
DSO	Distinguished Service Order
EATS	Empire Air Training Scheme
Hamp	Allied codename for Mitsubishi A6M3 (Model 32) Type 0 carrier fighter
Hap	Allied codename for Mitsubishi A6M3 (Model 32) Type 0 carrier fighter
Helen	Allied codename for Nakajima Ki-49 bomber
HMT	Hired Merchant Transport
HMS	His Majesty's Ship
IJA	Imperial Japanese Army
IJN	Imperial Japanese Navy
Jagdgeswader	German air force fighter wing
Jake	Allied codename for Aichi E13A reconnaissance floatplane
JG	Abbreviation of *Jagdgeswader*
Lily	Allied codename for Kawasaki Ki-48 light bomber
Luftwaffe	German Air Force
NCO	Non-Commissioned Officer
Nick	Allied codename for Kawasaki Ki-45 two-seat fighter
Oberst	*Luftwaffe* rank equivalent to colonel
Oscar	Allied codename for Nakajima Ki-43 fighter
OTU	Operational Training Unit
Pete	Allied codename for Mitsubishi F1M observation floatplane
POW	Prisoner of War
RAAF	Royal Australian Air Force
RAF	Royal Air Force
Ramrod	RAF codename for an operation similar to a Circus, but with the primary objective of destroying the ground target.
Regia Aeronautica	Royal Italian Air Force

Rhubarb	RAF codename for small scale attacks on ground or air targets.
Roadstead	RAF codename for low-level anti-shipping attacks by bombers escorted by fighters.
Rodeo	RAF codename for fighter sweep not involving bombers.
RSU	Repair and Salvage Unit
Rufe	Allied codename for Nakajima A6M2-N fighter floatplane
Sally	Allied codename for Mitsubishi Ki-21 bomber
Sentai	Abbreviation of *Hiko Sentai*, a Japanese Army flying regiment
TAF	Tactical Air Force
Tony	Allied codename for Kawasaki Ki-61 fighter
USAAF	United States Army Air Force
VC	Victoria Cross
VIP	Very Important Person
WWII	World War Two
Zeke	Allied codename for Mitsubishi A6M *Reisen* (Zero) Type 0 carrier fighter

MAPS

A map of the UK showing the different fighter sectors in WWII and the bases used by No. 452 Squadron. The unit was established in April 1941, and it was originally planned for it to be located at the newly constructed base at Valley, on the island of Anglesey. However, this plan was changed and the squadron was instead established at Kirton-in-Lindsey in Lincolnshire.

In July 1941 No. 452 Squadron moved forward to No. 11 Group where it operated from Kenley and Redhill in Surrey. For almost a year the squadron was keenly involved in combat with the Luftwaffe from these bases, often operating over the English Channel and northern France.

In March 1942 the squadron withdrew to RAF Station Andreas on the Isle of Man for a period of rest. It was then briefly at Baginton in Warwickshire, before it moved to Australia in June.

A full list of UK bases is on page 11.

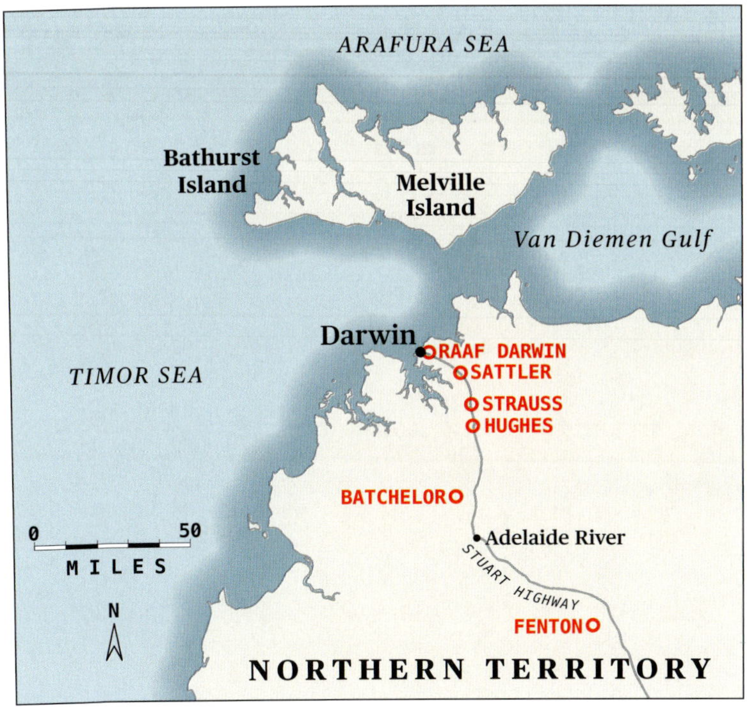

This map shows bases used by No. 452 Squadron in the Darwin area in 1943-44. After arriving in Melbourne in August 1942, the squadron spent several months training at Richmond and Bankstown near Sydney. In January 1943 it moved north to the Darwin area where it was briefly based at Batchelor before it settled into Strauss where it remained for over a year. In May 1944 the unit moved to nearby Sattler where it remained until advancing north to the Netherlands East Indies at the end of that year.

A full list of Pacific bases is on page 11.

This map of the Netherlands East Indies shows the key Allied base of Morotai where No. 452 Squadron was based for the first six months of 1945. In the final months of the war the squadron was involved in the Allied invasion of Borneo and operated from Tarakan and Balikpapan.

A full list of Pacific bases is on page 11.

BASES

No. 452 Squadron UK Bases 1941-1942

From	To	Base	Remarks
8 April 1941	21 July 1941	RAF Station Kirton-in-Lindsey, Lincolnshire	No. 12 Group
21 July 1941	21 October 1941	RAF Station Kenley, Surrey	No. 11 Group
21 October 1941	14 January 1942	RAF Station Redhill, Surrey	No. 11 Group
14 January 1942	23 March 1942	RAF Station Kenley, Surrey	No. 11 Group
23 March 1942	12 June 1942	RAF Station Andreas, Isle of Man	Ground echelon at RAF Atcham, Shropshire From 1 June 42
12 June 1942	18 June 1942	RAF Station Baginton, Warwickshire	No. 9 Group
18 June 1942	13 August 1942	en route to Melbourne, Australia	

No. 452 Squadron Pacific Bases 1942-1945

From	To	Base	Remarks
13 August 1942	6 September 1942	RAAF Station Laverton, Victoria	
6 September 1942	12 November 1942	RAAF Station Richmond, New South Wales	
12 November 1942	16 January 1943	RAAF Station Bankstown, New South Wales	
17 January 1943	1 February 1943	Batchelor Airfield, Northern Territory	No. 1 Fighter Wing
1 February 1943	8 March 1944	Strauss Airfield, Northern Territory	Dets. at Wyndham, Western Australia and Milingimbi Island, Northern Territory
9 March 1944	27 March 1944	Guildford Airfield, Perth, Western Australia	
27 March 1944	3 May 1944	Strauss Airfield, Northern Territory	
3 May 1944	11 December 1944	Sattler Airfield, Northern Territory	No. 80 Wing from June 1944
13 December 1944	29 June 1945	Pitoe, Morotai, Netherlands East Indies	Joined 1st TAF
29 June 1945	31 December 1945	Juwata Airfield, Tarakan Netherlands East Indies	Det. at Balikpapan airfield, Borneo
31 December 1945	-	Squadron disbanded	

No. 452 Squadron Bases Australia from 2010

From	To	Base	Remarks
1 December 2010	Current	Reformed RAAF Base Darwin, Northern Territory	Flights at Darwin, Tindal, Townsville, Amberley and Oakey

PREFACE AND ACKNOWLEDGMENTS

Of the nineteen RAAF squadrons that went to England between 1941 and 1945, most have passed into history. But in more recent times, a few have been reborn into the modern RAAF. No. 452 Squadron is one of them.

I set myself three aims with this publication which is why it is presented the way it is. First, I wanted the book to be a faithful record of the past squadron's life with all the stories, some anecdotes and a brush through the "official records" combining to tell the tale. Second, it had to be a record of service for all and a commemoration of the 49 members who gave their all. Third, and primarily, I wanted to present the story of what happened, so the families of squadron members and succeeding generations could understand what it was like and then present it in a way they might understand. I hope I have succeeded.

As reflected in the bibliography, extensive source material has been consulted. I have endeavoured to check the facts with at least two independent sources (usually three) and where items could not be substantiated, comment has been made. Wherever possible, I have mentioned squadron members by their full name, but it was just not possible to mention everyone. With significant numbers of officers and aircrew NCOs serving with No. 452 Squadron and over four times that many as ground crew, I trust the reader will understand my predicament. Having said that, omission does not infer the member did not contribute. Far from it, squadrons such as No. 452 worked as a team – aircrew and ground crew alike.

In any fighter squadron story, there is always the controversial matter of aces. Who, how many claims, what proof and so on. Consequently, I have deferred to the most thorough consideration of scores by British historians Christopher Shores et al, who are considered the world experts on British and Commonwealth claims. Their work is referenced in the bibliography.

I have also not gone into detail about the Darwin air raids of 1942-44, some of which No. 452 Squadron scrambled to intercept. There are dozens of books, articles and documentaries on the subject covering every detail, so again, the reader can look elsewhere should they wish for further information.

For me it has been an absorbing journey of discovery, the most rewarding aspect of which has been remembering what the squadron personnel did. I would like to pay credit to some of the principal agents who assisted. First and foremost, to the helpful staff at RAF Air Historical Branch in the UK, Ms Monica Walsh at the RAAF Museum and Ms Roz Turner at the RAAF's Office of Air Force History. Other contributions are credited in the bibliography.

As an historical record, I have used some conventions which some may find cumbersome, but I make no apologies for that. During the war, aircraft were each given a serial number at the factory and later, an alphabetical code letter was applied to the fuselage for visual identification and was used as the individual radio call sign. I have included these details for aircraft enthusiasts and scale modellers and used the convention "code letter: serial". For example, aircraft QY-A / BN537. The "QY" denotes an aircraft of No. 452 Squadron followed by the aircraft's individual code letter "A". Then follows the aircraft factory serial number as assigned by the RAF, "BN537". Place names used are spelt as they appeared at the time in the Operations Record Books (also called Unit History Sheets, RAAF Form A50 or RAF Form 540). Units of measurement have also been kept as imperial as those were used at the time. Photographs are captioned and credited to the copyright owner or source. Where there is no source mentioned, these are RAAF official from either the RAAF Museum or the Office of Air Force History.

Finally, I want to pay a special tribute to my wife Margaret who has once again had to put up with a rather distracted husband, at least for the ten years on and off that this took to produce.

I hope that you enjoy the No. 452 Squadron story.

Mark Lax
Canberra, Australia
January 2026

INTRODUCTION

The declaration of war by Britain on 3 September 1939 and Australia's rush to follow suit a day later caught both country's defences sadly lacking and ill-prepared for the coming conflict. For all three services, this meant mobilising from a peacetime cadre to rapidly building both defensive and offensive forces. Four days later, the British government declared that the main weakness of the Allies *vis-à-vis* Germany was in air strength - and this included any contribution from the young and struggling Royal Australian Air Force.

Formed in March 1921 from the remnants of the Australian Flying Corps, the RAAF then was little more than a small training organisation with obsolete World War I aircraft. Subsequent government funding for the new service had been so paltry that up to 1936 it was only enough for the RAAF to maintain just two composite squadrons, a seaplane flight, a flying training school and three rudimentary air bases. By 1937, several damning reviews forced the government's hand but by the end of 1939, the RAAF still only had a total personnel strength of 3,489. To meet the demands of war, 246 aircraft formed the front-line strength of which only 164 were operational machines. At that time, the RAAF had three squadrons in Melbourne, four squadrons in Sydney and two at Perth. Darwin and Brisbane had a squadron apiece.

Hence by late 1939 the RAAF had twelve squadrons, yet only seventeen days after the declaration of war, Australia generously offered six of these squadrons for service with Britain. Initially, this proposal was left in abeyance and apart from the men of No. 10 Squadron already in England, another 450 Australians (of whom at least 159 were pilots) were serving with the RAF as part of a pre-war training arrangement. No. 10 Squadron had been sent to England in 1939 to take delivery of Short Sunderland flying boats then on order. Australia offered the squadron's services for maritime patrol and No. 10 Squadron remained in Britain for the remainder of the war. It was just the beginning of what would become a substantial effort.

Other RAAF squadrons were to follow with overseas deployment. By 1940, No. 3 Squadron had been despatched to the Middle East and Nos. 1 and 8 Squadrons to Malaya. By early 1942, the RAAF was in action on four fronts: Europe, the Atlantic, the Middle East and in the Asia-Pacific. When the European war ended, in May 1945, 15,000 Australians had been employed in the air war against Germany and Italy. In all, 6,636 RAAF men died in action in the European theatre, one-fifth of the entire Australian war deaths from all services and campaigns of the war.

There can be little doubt that the RAAF came of age as a fighting force during World War II. Although arguably the world's second oldest air force, it was little more than a flying club until the war forced its rapid expansion and technical development. At the height of the war, the service had expanded to a strength of 20,000 officers, 144,000 airmen and 18,000 airwomen. Over the war years, the RAAF grew to 50 squadrons in the Pacific and seventeen in Europe - a remarkable achievement. It had over 3,000 operational aircraft and nearly 3,000 trainers. RAAF members served in every theatre of operations and by the war's end the RAAF was considered the fourth largest air force in the world.

To achieve these amazing statistics from a country whose wartime population was around seven million, required sound organisation and willing volunteers. Australia like Canada, New Zealand and Rhodesia contributed men to a new scheme - the Empire Air Training Scheme (EATS) - a massive training organisation set up to provide Britain with the men it badly needed to fly and fight. The EATS is a remarkable story on its own, but not for these pages. However, under the scheme, the RAAF also promised eighteen fighting squadrons, known as Article XV squadrons from the respective article in the EATS charter. Seventeen eventually formed, and these saw service in Europe, the Middle East and the Pacific.

This is the story of one of those squadrons.

RAAF Hawker Demon biplanes in the 1930s. A tiny force prior to WWII, the RAAF would see extraordinary growth during the war. (Author collection)

An overhead view of RAF Kirton-in-Lindsey in Lincolnshire, which had grass runways. No. 452 Squadron formed at this location in April 1941. (National Transport Trust UK)

CHAPTER 1

APRIL TO JULY 1941 · HUMBLE BEGINNINGS

At the outbreak of World War II, the British government quickly realised it did not have enough men or materiel to fight the German *Luftwaffe* and Italian *Regia Aeronautica* on their own. Aircraft production could be ramped up at home, but men would have to come from elsewhere to satisfy the demand for air and ground crews. To solve the problem, the British government put forward a plan to train men from across the empire, and the challenge was accepted by the Australian, Canadian and New Zealand governments. The plan, signed on 17 December 1939, would be called the Empire Air Training Scheme and under the terms of its Article XV, Australia agreed to provide 28,000 trained aircrew over three years.

The proposal entailed a massive enlargement of the RAF by twenty per cent, with these numbers coming from across the Commonwealth. In the 1930s, the RAAF had trained fifty pilots who would be sent to serve in the RAF to gain experience, and in the event of war, revert to RAAF status. As well as these, at the beginning of the war, numerous Australians joined the RAF directly on either permanent or short-service commissions or as ground trades. Others living in Britain enlisted directly so by September 1939, about 450 Australians held RAF appointments scattered across the force. Some members would later join RAAF squadrons as the war progressed. The RAAF's commitment to the EATS scheme was for eighteen squadrons and of these, six were to be fighter squadrons. No. 452 Squadron would be the third fighter squadron raised in this way.

As agreed by the four signatories, individuals recruited into the RAAF would remain in their parent service and continue to wear their RAAF uniform, but after training, would be posted to fill RAF billets. The RAF would accept full financial responsibility except where Australian rates of pay and conditions differed from the British. Australia then made up the difference.

By the end of 1940, the worst of the Battle of Britain was over and Hitler's planned invasion of England, *Operation Sealion*, had been postponed if not cancelled. Although the British continued to suffer air raids well into 1941, these became more sporadic, easing the daily stress of the *Luftwaffe* onslaught on RAF Fighter Command, and none the least, the British populace. As well as bombing British towns and industrial centres, the *Luftwaffe* continued to mount night intruder raids which regularly claimed RAF bombers and fighters, usually as they approached to land – a time when they were most vulnerable.

The European air war had reached a small but significant milestone on 10 January 1941, when the RAF dispatched an escorted bomber force to bomb a German ammunition depot just south of Calais in the Forêt de Guines. This was the first time since mid-1940 that bombers had been escorted beyond the French coast. This was *Circus 1* – six Blenheim light bombers of No. 114 Squadron escorted by no less than eleven squadrons of fighters. There was light enemy opposition and all the Blenheims returned safely, with the squadron earning a congratulatory message from the prime minister. A "Circus" was the code name for a broad sweep over enemy territory designed to bring the enemy fighters into the air where they could be engaged. In time, *Circus* operations would become No. 452 Squadron's mainstay.

A *Circus* operation was air power intensive. The "Circus" was a box, normally of six light bombers, which although they bombed the allotted tactical target, were primarily designed to cause an enemy fighter response so they could be engaged by the RAF's fighter force. The bombers were escorted to the target by two fighter wings each of three squadrons, one wing acting as close escort and the other as high cover. Near the target a target-support wing would join company to draw enemy attacks away, and if opposition was weak or absent it would make strafing attacks on the target itself. On the way home a rear-support wing would take over prime responsibility for defeating any enemy pursuers. Thus, a usual force of four wings (twelve fighter squadrons) was engaged in every *Circus* while additional squadrons might patrol the route shortly before the main formation, and "mopping-up" squadrons would sweep the area immediately afterwards to catch any German fighters still in the air.

As happened to several new RAAF squadrons in the early days of the war, No. 452 Squadron was to have two "beginnings". The RAF had promised that No. 452 would be an Australian fighter squadron manned by Australians with aircraft provided by the British. While the RAAF got its act together in Australia to send the men, the RAF decided to form the squadron peremptorily in April 1941 while they awaited the RAAF members to arrive. Under the agreement between the RAF and RAAF, No. 452 Squadron would comprise a headquarters flight, two flights ("A" and "B") of pilots, a servicing echelon for maintenance and equipment stores and a reinforcement pool. The squadron would comprise seventeen officers and 241 other ranks. The commanding officer would be of squadron leader rank, as were the commanding officers of other RAF fighter squadrons, and the most senior NCO would be a flight sergeant.

The original plan was for the squadron to form up at an airfield near Rhosneigr, a small village in Anglesey, North Wales. Today, the airfield is RAF Valley and a large training base, but then it was about as small and remote a site on the British Isles as you could get. The intent was to provide the new squadron with the opportunity to train up before it became operational, far from the enemy threat.

But with the urgent need to replace and rebuild fighter squadrons after the Battle of Britain, the squadron's location was quickly changed to RAF Kirton-in-Lindsey in Lincolnshire, about sixteen miles north of Lincoln. The base was in No. 12 Group, a larger parent RAF organisation responsible for fighter operations from central England – mainly Cambridgeshire, Lincolnshire and Yorkshire. Kirton-in-Lindsey was established during World War I after which it closed but it reopened as an RAF station in the 1930s.

The Kirton-in-Lindsey airfield comprised two grass runways of 1,000 yards length and was one of numerous No. 12 Group fighter stations protecting the eastern approaches to Lincolnshire and Yorkshire. The airfield had been originally developed for home defence and saw a gradual resurgence with the RAF build-up during the 1930s. Kirton had been host to several fighter squadrons since it re-opened for operations in April 1940. It would be home to No. 452 Squadron until 21 July 1941. In 1942, the station was relinquished to the USAAF's 8th Air Force as a fighter airfield to support their bombing operations. Kirton was returned to the RAF in May 1943 for a fighter Operational Training Unit.

Upon its arrival, No. 452 Squadron took over four rooms in the barracks block and was allocated Number 3 hangar where they would share the facilities with Nos. 65 and 121 Squadrons, RAF, which were also flying early model Mk I and II Spitfires. By 10 May, twenty RAAF pilots had arrived to join their RAF brethren, the latter being more experienced fighter pilots who would bring the Australians up to speed.

Although the Battle of Britain was officially over at the end of October 1940, the *Luftwaffe* was by no means a spent force in the west and devastating bombing raids across Britain continued well into 1942. While the RAF awaited the Australian airmen, they posted experienced British fighter pilots into the nascent squadron to get it operational as quickly as possible. No. 452 would be a Spitfire squadron and remained so throughout the war. The squadron was initially established with sixteen Spitfire Mk Is with two in reserve, all war-weary, early-model hand-me-downs used for training purposes, and one Miles Magister (serial V1083), a two-seat elementary trainer which could be used for general flying. With the arrival of the personnel, No. 452 Squadron officially formed on 8 April 1941 and set to work.

The Spitfire needs little introduction suffice to say that the Mk I and Mk II were similar in performance and by mid-1941, were already outclassed by the German Messerschmitt Bf 109E/F models. Consequently, they were relegated to the training role, but because of the demand for the higher performing Mk V Spitfires elsewhere, the older model Spitfires were, out of necessity, used on operations well into 1941. No. 452 Squadron would have to wait until August before their Mk Vs arrived.

To lead the squadron the RAF appointed Squadron Leader Roy Dutton, an RAF officer and veteran of the Battle of Britain. He was also one of their highest scoring aces at that time with over thirteen enemy aircraft destroyed. Dutton, born in Ceylon, had joined the RAF on a short service commission in 1936 and by April 1940 he was a flight commander of No. 145 Squadron. His score rose steadily and on 31 May, he was awarded a DFC for his courage in the air and a bar followed on 20 August. In April 1941, he was posted to Kirton-in-Lindsey to form No. 452 Squadron as its inaugural commanding officer. He would remain as commanding officer only briefly until 15 June 1941 when another veteran of the Battle of Britain and an Australian, Squadron Leader Bob Bungey, took over. From that time, the squadron could be called predominantly Australian in both composition and character.

Squadron Leader Roy Dutton, RAF, who was the initial commander of No. 452 Squadron.

Squadron Leader Bob Bungey, the first Australian commander of No. 452 Squadron.

Squadron Leader Bob Bungey

Bob Bungey was born in Fullarton, South Australia, in October 1914, and while he trained as a pilot with the RAAF in Australia, he transferred to the RAF in 1937. He was one of numerous RAAF pilots to do so – an arrangement struck with RAF authorities for fledgling pilots to gain experience. Bungey saw action in France and during the Battle of Britain with No. 145 Squadron flying Hurricanes. He was awarded a DFC on 7 October 1940, but on 7 November, he was shot down by Bf 109s and baled out. He returned to the squadron, but on 10 June 1941, he was posted to command No. 452 Squadron. Bungey's time as commanding officer was eventful as during this period, the squadron matured into an effective fighting unit. He shot down another Bf 109 on 6 December 1941, but in January 1942 he was posted out. After a knee operation he was repatriated to Australia and returned to Adelaide to be with his wife and young son. After she died suddenly from meningitis in May 1943, Bungey was deeply troubled and he took his own life with a pistol on 10 June, aged 28. His son who he also shot miraculously survived.

Meanwhile, the RAF filled the flight commander positions with experience as well. On 13 April 1941, an Irishman in the RAF, Flight Lieutenant Brendan "Paddy" Finucane, was posted in as commander of A Flight. Then within two days he shot down a Bf 109 over Dover to bring his score to six. Finucane was already a veteran Spitfire pilot and rising ace. The B Flight commander was another RAF veteran, Flight Lieutenant Alfred "Pinky" Douglas, who remained at No. 452 Squadron only until taking command of a Canadian squadron in September 1941.

Wing Commander Brendan "Paddy" Finucane

Wing Commander "Paddy" Finucane (pronounced Fin-oo-can) was born in Dublin in October 1920 and joined the RAF in May 1938. On graduation as a pilot, he was posted to No. 65 Squadron flying Spitfires. Finucane took part in the Battle of Britain, claiming five enemy aircraft destroyed and earning a DFC. In May 1941 he joined No. 452 Squadron as a flight commander. By late October, he had been awarded two more DFCs and a DSO. In mid-October, he broke an ankle scaling a wall during a blackout and was off operations until January 1942 when he was posted to sister Spitfire unit, No. 602 Squadron, RAF. By June 1942, he was appointed wing leader at Hornchurch. Finucane had added another seventeen to his tally before being killed when he ditched in the English Channel on 15 July 1942 after his Spitfire Mk Vb BM308 was hit by ground fire. While over the channel, the aircraft's engine seized, and he crashed into the water. It rapidly sank taking Finucane down with it. His final tally was 26 aircraft destroyed and six shared destroyed, eight probables and one shared probable, and eight damaged. He was just 21 years old.

Flight Lieutenant Brendan "Paddy" Finucane, RAF, in the cockpit of a Spitfire carrying the No. 452 Squadron code "UD". Note the shamrock symbol, celebrating his Irish heritage. (Imperial War Museum)

Flight Lieutenant Alfred "Pinky" Douglas

Finucane's fellow No. 452 Squadron flight commander was a Welshman from Colwyn Bay, Alfred Douglas. Known as "Pinky", he had been born in 1908. He joined the RAF Reserve in 1928 as a part-timer. He spent the 1930s and early war years in a training role. By Spring 1941 as a flight lieutenant, he was posted to No. 74 Squadron, RAF, as a flight commander to fly Spitfires before transferring to No. 452 Squadron. After spending only a few months at No. 452, he went on to command No. 403 (RCAF) Squadron and then No. 401 (RCAF) Squadron. By May 1942, he was commanding RAF Station Hawkinge and by war's end, he was commanding RAF Station Warmwell.

Joining these experienced officers, RAF Flight Sergeants Cooper and Chadwick were posted in from No. 71 (Eagle) Squadron as the senior ground staff in charge of A and B Flights respectively. The Eagle squadrons were comprised mainly of Americans who had joined the RAF in search of adventure, not waiting to see when or if America would join the war. Within days, more airmen were posted into No. 452 Squadron from various other units, and by 11 April 1941, enough were present to fill both flights. These airmen were mostly British, but in time, this would quickly change. As if to welcome them into the war, on 17 April, the *Luftwaffe* paid the squadron a visit and dropped a stick of bombs across the aerodrome, but little damage was done.

Negotiations to construct the EATS agreement were commenced in October 1939 and while each of the dominions pledged aircrew to Britain, the understanding was that squadrons with a clear national character would be maintained. Australia's contribution would be the numbered squadrons 450 to 467 inclusive. Although it took almost a year for the agreement to be signed, Australians embraced the commitment, and many young men were soon to find themselves members of a rapidly expanding all-volunteer force.

Indeed, the idea of joining the air force was popular with Australia's youth and while almost all wanted to be pilots, the majority would end up on the ground. Men with trade qualifications were most welcome and such was the enthusiasm in Australia that a rapidly mounting surplus was becoming embarrassing to authorities with nowhere to send them. To overcome the situation, on 25 March 1941, the RAAF's liaison officer in London, Group Captain Frank McNamara, VC, was instructed to offer 1,000 of these men to the RAF. This offer was quickly accepted and after some haggling as to their destination, about 600 were sent to UK while the rest went to the Middle East. The need was pressing as by mid-1941, RAF Fighter Command comprised 84 squadrons, all needing engine and airframe fitters as well as electrical and mechanical trades to keep the fighters serviceable.

Friday 23 May 1941 was just another normal working day at RAAF Williamtown, a rudimentary RAAF airfield just north of Newcastle, New South Wales. Early that morning an administrative officer, Flying Officer Bill Crossing, reported to the Base Orderly Room for duty. He was there to form the nucleus of a new fighting squadron, No. 452, and had been appointed as the unit's temporary commanding officer. He would also serve as the squadron adjutant.

On the following Monday, he was joined by Flight Lieutenant Bill Seldon, the unit medical officer, and within days airmen of all musterings began to arrive. Meanwhile the Williamtown station commander, Squadron Leader John Paget, held the overall responsibility for bringing the ground crew component of several new Article XV EATS units up to scratch. Together with Crossing, he ensured the airmen continued their general service training. Between May and July six such units were in varying stages of formation because of McNamara's offer, and all were at Williamtown. By the end of June No. 452 Squadron had grown to 158 other ranks, together with the two officers Crossing and Seldon.

For these men at Williamtown, Article XV meant little. Their days consisted of drill, general service training and organised sport. On the weekends, the men often attended dances in the Newcastle town hall and concerts and movies in the base recreation hut. Leave too, was a regular event. For them, the war must have seemed a long way away. Aircraftman Jim Grant recalled his time at the Williamtown base during this period:

> I never anticipated that ten weeks of boredom and frustration would pass before my genuine final embarkation leave would be taken. After all, my colleagues and I had volunteered to join the RAF Infiltration Scheme, had been granted embarkation leave and therefore expected we would soon be on our way … At the time I was there, Williamtown was little more than a landing strip, with a few buildings, mostly of a temporary nature, stuck in an inhospitable bit of coastal scrub. The roads through the camp were lined with railway sleepers, so at least you could tell where the roadside finished, and the sand dunes and scrub commenced. One of our regular jobs, when all else failed was to whitewash the sleepers.

While at the time Grant was a member of another Article XV unit, No. 454 Squadron, he would eventually serve with No. 452's sister Spitfire squadron, No. 457, in Darwin.

By early June, bayonet and gas drills, kit inspections and rifle practice had begun, but there was no technical follow-on training and nothing to prepare the men for their eventual

Airmen present their kit for inspection at RAAF Williamtown, circa 1941. (Jim Grant)

destination. But the members of No. 452 Squadron did not have long to wait for a move. On Wednesday 11 June, orders came through for overseas movement and the squadron quickly packed and entrained for embarkation from Sydney on HMT *Awatea* two days later.

Meanwhile in the UK by the end of April, RAAF pilots had trickled into Kirton-in-Lindsey from other RAF units and No. 57 OTU, a Spitfire training unit. Among the new arrivals were Pilot Officer Ray Thorold-Smith and ten RAAF sergeant pilots. Thorold-Smith, known as "Throttle", was to become one of the squadron's stalwarts and would later take the squadron back to Australia. Nearly all of the RAAF pilots had trained at Narromine, before passing through a Service Flying Training School in Canada followed by an Operational Training Unit in the UK. By the middle of April, twenty RAAF pilots had arrived and a "vigorous training program was initiated, as most of the Australians had no previous squadron or operational experience".

Even experienced pilots make mistakes. While carrying out an air display near Halifax on 3 May, Finucane's aircraft hit Dutton's and despite partially severing the empennage, Dutton managed to fly the aircraft into a controlled crash landing. He ended up going through a stone wall. The unit historian was to write:

> The CO then received medical attention at a hospital in Halifax, later returning to Kirton-in-Lindsey by transport. Flight Lieutenant Finucane, however, was able to carry on.

By mid-May, slightly better performing Spitfire Mk IIs were collected from No. 303 (Polish) Squadron at Northolt as they were converting to the Mk V version. As such, the unique fuselage code letters of "UD" assigned to No. 452 Squadron would be a while forthcoming and some original photos show the squadron Spitfires still carrying previous unit codes of "RF" – a cause of confusion for some later air historians.

On 22 May, No. 452 Squadron was "at readiness" for the first time, but although technically operational, little action was forthcoming with the unit flying convoy and defensive patrols until July. By 2 June, the RAF declared No. 452 Squadron officially "day operational" and placed them on the daily readiness schedule. The war was now "on".

With things warming up, and given inexperience, accidents inevitably happened. On 4 June Thorold-Smith collided with a Hurricane of No. 56 OTU on take-off, writing off the Spitfire but not himself. Two weeks later, Sergeant Raife Cowan crashed his Spitfire and suffered a fractured elbow.

Between 22 May and 11 July, during what was called by the official historian "an almost unbroken spell of good weather", 189 operational patrols were carried out, all without incident. By the end of July, the squadron was taking part in *Rhubarb* and *Roadstead* operations over France and Belgium. *Rhubarbs* were small scale attacks on ground targets, usually in France while *Roadsteads* were bomber escorts on low-level anti-shipping attacks. Throwing them into the fray was all part of bringing the squadron up to speed as quickly as possible. However, it would be *Circus* operations designed to draw up enemy fighters to be engaged that would get most squadron attention in the coming months.

The underside of a Messerschmidtt Bf 109 often incorrectly referred to as an ME 109. (Author collection)

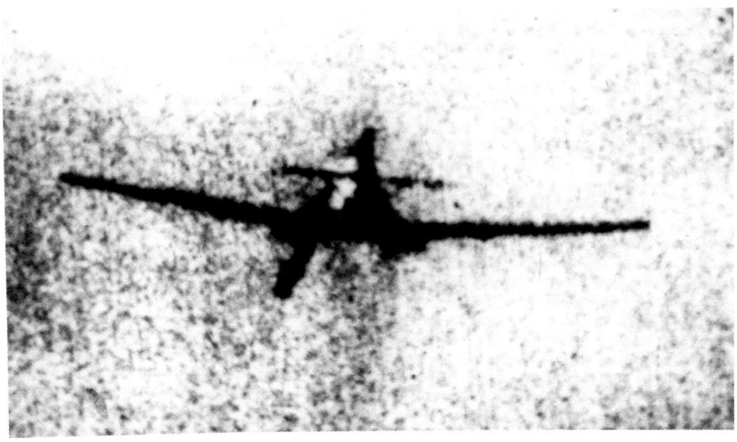

Grainy gun camera footage of a Bf 109. (Author collection)

Finally, on 11 July, using West Malling as a forward operating base, No. 452 Squadron took part in *Circus 44* over France together with Spitfires from Nos. 92, 485 and 609 Squadrons making up the formation. Indeed, some 108 Spitfires from nine squadrons took part in the *Circus*. The intent was that any ensuing melee would occupy the *Luftwaffe*, thus letting a low-level bombing attack against the Lille railway yards get through to the target. The collective squadrons rendezvoused over Manston at 1435 and crossed the French coast ten minutes later. The wing then split with the Australians at 18,000 feet and behind the main gaggle when they were attacked by eight Bf 109s of *Jagdgeswader* 26. It was during this engagement that Finucane in Spitfire P8038 shot down one of them – No. 452 Squadron's first victory. Having sighted the enemy approaching, he followed the Messerschmitt into a dive and shot it down at close range. Being slightly to the rear of the main force, none of the other Australians were in position to attack, so they returned to base. These *Circus* tactics were to prove problematic and often resulted in as many losses as claims.

This was a tit-for-tat engagement, and Sergeant Alex Roberts flying Spitfire P7562 failed to return from this operation but survived. He later recalled his adventure for British intelligence and recorded his memoirs in a diary, now held in the Australian War Memorial collection. He recalled:

> We had taken off from West Malling … and the plan to draw the enemy seemed to be working because we were getting plenty of attention from ME-109s. One of these got a burst of cannon shells into the tail unit of my Spit. The little plane shuddered and went into a dizzy spiral dive. Speed was building up and I had no control, so I thought it was time to bale out. I got out of the Spitfire at 22,000 feet, worrying at first about the lack of oxygen at that height, deciding I should free-fall a few thousand feet. With that behind me, I pulled the ripcord, praying that the chute would work. It did. My fall was stopped with a gut-wrenching jolt, and I dangled there in the sudden quiet until the attacking ME-109 interrupted things.

After landing in France near Calais, Roberts was unhurt, but his first worry was being machine gunned by the Messerschmitt that shot him down. Having survived a rough landing, he was to have a remarkable escape and evasion. He landed just outside the perimeter fence of the German fighter base at St Omer and while the guards scrambled to get to him, he was whisked away by a Frenchman with the Resistance. After some time on the run, he made it as far as the border of unoccupied France but was eventually betrayed and arrested by French Gendarmes. After questioning, he was handed over to the Germans and sent off to a concentration camp. Three weeks later, he escaped and with eight others, crossed the Pyrenees and returned to the UK to be posted back to No. 452 Squadron where he flew several sorties before being posted to No. 258 Squadron, RAF. This unit would serve in the Far East, there being a new RAF policy not to return escapers to the same theatre of operations. Had he been shot down again and caught, he would most likely have been executed by the Gestapo.

A portrait of Sergeant Alex Roberts, who became a POW in France but escaped and returned to the UK. (Alex Roberts)

There had been numerous accidents and "prangs" during the first few months leaving a few broken bones and much pride damaged, but on an operational flight over Lincolnshire on 5 July, Sergeant Andrew Costello was to be the first squadron fatality when his Spitfire P8085 was intercepted by a German Ju 88 night intruder of NJG/2, and shot down over Fen Lane Farm just after 0100. He was coming into land at North Coates aerodrome and probably did not know what hit him.

After the Battle of Britain was technically over, the RAF set upon a course of taking the fight to the enemy, and that meant cross-Channel excursions. Despite their lack of experience,

Sixteen No. 452 Squadron pilots in June 1941 during the early days of operations. Left to right: Paul Makin, Ken Cox, Arch Stuart, Brendan Finucane, Ian Milne, Jim Hanigan, Fred McCann, Ray Thorold-Smith, Bob Bungey, Bill Willis, Alex Roberts, Don Lewis, Graham Douglas, Andy Costello, Keith Truscott and Dick Gazzard.

there was little time for members of No. 452 Squadron to comfortably settle into their role as RAF Fighter Command commenced their offensive. With this came the first move. On 21 July 1941, the squadron received orders at short notice to move to RAF Station Kenley in Surrey and transfer to No. 11 Group. Kenley was established in 1917 and remained an important base until operations ceased there in 1959. Located just south of Croydon, the station was one of three main Battle of Britain airfields in No. 11 Group, RAF Fighter Command. Comprising two 1,200-yard asphalt runways and good accommodation, the move was a welcome change. No. 11 Group was closer to the action and the squadron's role would now go from the defensive to the offensive, all controlled from the fighter control room at Uxbridge. No. 452 Squadron would join the Kenley Wing and soon find itself flying large formations with fellow squadron Spitfires of Nos. 485 (NZ) and 602 Squadrons. Operations began the next day and soon the squadron was carrying out sweeps over the Channel which would be flown on almost a daily basis, weather permitting.

Meanwhile, the Germans had stationed the battle cruisers *Scharnhorst* and *Gneisenau* and the heavy cruiser *Prince Eugen* at Brest on the channel coast, so the RAF had been planning an operation against these capital ships for some time. Once under sail, any of these cruisers could do enormous damage to Allied shipping, so they were deemed a significant threat. As the final stages of planning for an attack on Brest were underway, the Germans quietly sailed the battle cruisers down the coast to La Palice forcing a quick rethink.

The RAF planned to intercept and attack the vessels as they moved north up the coast. While the main diversionary effort on 24 July would be against the harbour facilities at Brest and La Palice, a further diversion was planned for Cherbourg to draw the *Luftwaffe* defences away. Thus, *Operation Sunrise I* and later that day, *Sunrise II*, was planned with eighteen Blenheims each as the diversionary bomber force. The ruse failed and although some damage was done to the Cherbourg docks, no enemy aircraft appeared and no aircraft were lost. The RAF would have to wait until February 1942 before having another go.

With perhaps some consolation, when No. 452 Squadron took part in *Sunrise II*, Flying Officer Andrew Humphrey in Spitfire Mk IIa P7973 shot down a Messerschmitt Bf 109E on the squadron's return journey. Humphrey would go on to become an ace with seven destroyed and two probables and would later rise to the highest possible rank, that of Marshal of the RAF. Today he is remembered by the Humphrey Medal presented to the Dux of the RAF General Duties Aerosystems Course held annually at RAF Cranwell. Humphrey completed his brief tour with No. 452 after spending only a month with the squadron before being posted to a training unit as an instructor.

Meanwhile, fresh pilots continued to trickle in. Two such Australian newcomers illustrate the varied paths into the squadron ranks, both arriving on 25 July. The first was Sergeant Charles Manning. Manning had enlisted in Melbourne in July 1940 and after passing through No. 2 Initial Training School at Bradfield Park, he underwent his basic pilot training at No. 4 Elementary Flying Training School at Mascot. He was then sent to Canada for further experience at No. 1 Service Flying Training School, Camp Borden, Ontario, and finally to England and on to No. 53 Operational Training Unit. He remained at No. 452 Squadron for just under two months.

The second newcomer, Sergeant Bruce Dunstan from Chelsea in Victoria, followed a different path. He enlisted in the RAF in England. After initial training, he underwent a pilot's course at No. 13 Elementary Flying Training School at White Waltham. Then followed No. 7 Service Flying Training School at Montrose and finally on to Spitfire conversion at No. 53 OTU at Heston and Llandow. Dunstan's time at No. 452 Squadron was short but action filled. On 18 November, he was shot down and was rescued after baling out into the sea. Dunstan was posted to No. 41 Squadron, RAF, to fly Spitfires in December after just five months at No. 452.

Both of these pilots would not survive the European air war. Charles Manning was shot down and killed on 18 September 1941 and Bruce Dunstan on 12 February 1942. They were aged just twenty and twenty-one respectively.

Spitfire Mk II P7973 as flown by Squadron Leader Keith "Bluey" Truscott. The code letters "RH" are supposedly the codes of the RAF Central Gunnery School but may also refer to "red head" in reference to Truscott's hair. The aircraft is now in the Australian War Memorial.

No. 452 Squadron Spitfires at Kirton-in-Lindsey airfield prior to the July 1941 move to Kenley.

A gathering of No. 452 Squadron pilots pose for a photograph. Left to right are Sergeant Jack Donald, Sergeant Raife Cowan, Flight Lieutenant Brendan "Paddy" Finucane, Sergeant Ed Jackson and Sergeant Eric Schrader.

CHAPTER 2

AUGUST TO DECEMBER 1941 · SWEEPING THE SKIES

The warm summer weather of August 1941 heralded a busy month as No. 452 Squadron continued its regular missions including *Circus* operations with great success. On 3 August, No. 452 led No. 602 Squadron on an evening sweep of the Gravelines-St Omer-Ambleteuse area when they were attacked by five Bf 109s of *Jagdgeschwader 26* – seasoned fighter pilots. Finucane claimed one destroyed and a probable while Pilot Officer Bill Eccleton claimed another destroyed. Other individual combats were inconclusive and there were no friendly losses.

Early in the month, the Australian contingent which had sailed in June finally arrived in the UK. Flight Lieutenant William Seldon became the unit's medical officer and Pilot Officer Harry White served as a temporary squadron adjutant while Crossing was away on a course. Thirty-four airmen swelled the ranks of the ground crew and technical trades. As well as fresh personnel arrivals, the squadron also began to re-equip with the Spitfire Mk Vb, a much superior version of the marque now fitted with two 20mm cannon and four machine guns, for a much-needed boost in firepower.

Spitfire Mk Vb and Messerschmitt Bf 109F

By 1941 steady improvements by designers had increased the performance and lethality of fighter aircraft, particularly in the hands of an ace. The Spitfire Mk V was an example. Powered by a 1,470hp supercharged Merlin engine, it had a maximum speed of 370mph at 20,000 feet, the usual cruising altitude. By comparison, the Bf 109F had a Daimler-Benz engine developing 1,300 hp giving the Messerschmitt a speed of 388mph at the same altitude. The Spitfire could turn more tightly while the Messerschmitt had a faster rate of climb, making both aircraft a reasonably even match, except in a close dogfight.

However, it was the Mk Vb armament that greatly improved the Spitfire's firepower. The eight 0.303-inch calibre Browning machine-guns that the Spitfire Mk II carried were replaced by two 20mm cannon and four 0.303-inch Browning machine guns giving the Mk Vb Spitfire much more hitting power. Messerschmitt had already learned the benefit of cannons from their Spanish Civil War experience and even the earliest Bf 109 models carried one. The Bf 109F now carried a single 20mm cannon in the nose and two 7.9mm machine guns. Hence both aircraft had considerable hitting power.

Spitfire Mk II, Spitfire Mk Vb and Messerschmitt Bf 109F Technical Data

	Spitfire Mk II	Spitfire Mk Vb	Messerschmitt Bf 109F
Power Plant	1,175hp Merlin	1,470hp Merlin	1,300 hp Daimler Benz 601N
Dimensions			
Span	36 feet 10 inches	36 feet 10 inches	32 feet 6 inches
Length	29 feet 11 inches	29 feet 11 inches	29 feet
Height	11 feet 5 inches	11 feet 5 inches	11 feet 6 inches
Weights			
Empty	4,900 lbs	5,065 lbs	5,269 lbs
Loaded	5,900 lbs	6,622 lbs	6,393 lbs
Max	6,317 lbs	6,700 lbs	6,872 lbs
Performance			
Max Speed	357 mph	371 mph	388 mph
Cruising Speed	195 mph	226 mph	348 mph
Rate of Climb	2,995 ft/min	3,250 ft/min	4,290 ft/min
Service Ceiling	33,900 ft	37,000 ft	37,700 ft
Normal Range	-	470 miles	-
Max Range	500 miles	1,170 miles	410 miles
Armament	IIA – 8 x 0.303-inch MG IIB – 2 x 20mm cannon & 4 x 0.303-inch MG	2 x 20mm cannon & 4 x 0.303-inch MG	1 x 20mm cannon & 2 x 7.9mm MG

Spitfire Mk Vb AB852 (UD-W) as flown by "Paddy" Finucane. Note the shamrock forward of the cockpit.

By now, No. 452 Squadron's pilots were beginning to perform like experienced fighter pilots. The prangs of the early days had virtually ceased, and the squadron's tally began to steadily rise. So too did its losses. It was during *Circus 68* against Gosnay power station as the target that on 9 August five Bf 109s were shot down, for three own losses. Finucane as usual was in the thick of it, as were the Australians. Flying at 20,000 feet over the Mardyck-Bethune area, Red and Yellow Sections encountered about thirty Bf 109s of JG 26 led by their

Finucane's Spitfire Mk Vb AB852 (UD-W) being loaded with belts of 0.303-inch rounds by armourers. The serial has been erased by the censors.

commanding officer, the redoubtable *Oberst* Adolf Galland. They were in for a fight. A large, ever-expanding dogfight ensued which spread from the target area to the coast. Finucane claimed one and two shared, the first with Ray "Throttle" Thorold-Smith and the other with Sergeant Keith Chisholm. Chisholm then joined Pilot Officer Don Lewis to claim the fourth, while Pilot Officer Keith Truscott claimed the fifth. Lewis later recalled what happened for the intelligence officer:

> I was White Two of centre section of squadron formation. I sighted an ME109 approaching on starboard side. I attacked and fired a burst of about three seconds. Attack was from beam to quarter. The e/a [enemy aircraft] went into a shallow dive and the pilot baled out. I understand that Yellow Two, who was behind me at the time, fired at e/a also. I attempted to rejoin formation but two ME109's attacked me. They climbed on my starboard, but I could not get my sights on them as my speed was too slow. I then dived to cloud at 6,000 feet. When this cloud layer finished, I dived into broken cloud at about 1,500 feet.

Note that Yellow Two mentioned above was in fact Red One – Finucane.

However, Pilot Officer Justin O'Byrne together with Sergeants Geoff Chapman and Gerry Haydon did not return. O'Byrne would see out the war in *Stalag Luft III*, while the other two were killed. Pilot Officer Keith "Bluey" Truscott, so nicknamed because of his fiery red hair, claimed his first that day while Sergeant Keith Chisholm claimed two half-shares. There would be much more from Truscott.

Squadron Leader Keith "Bluey" Truscott

Pilots Flight Lieutenant Keith "Bluey" Truscott (left) and Sergeant Keith Chisholm are debriefed by Pilot Officer Denys Lane-Walters, No. 452 Squadron's RAF intelligence officer.

Keith Truscott was another Australian who had arrived in April 1941, and one that would quickly make his mark. Keith "Bluey" Truscott was born in Melbourne on 17 May 1916. Educated at Melbourne High School, he was regarded as an outstanding all-round sportsman and was captain of both the football and cricket teams. After school he took a job as a clerk at a local firm and played football for the Melbourne Football Club. He joined the RAAF in July 1940 and after training in Canada he was posted to England where he joined No. 452 Squadron. By the time he left the squadron to return to Australia in March 1942, he was a household name – a fourteen-victory ace with a DFC and bar and a Mention-in-Dispatches. He was recalled to Australia in March 1942 to join No. 76 Squadron in New Guinea for the defence of Milne Bay. With the Japanese threat passed, the squadron returned to Darwin where Truscott claimed a Betty bomber shot down, his only Japanese aircraft claim. Sadly, on 28 March 1943, while conducting manoeuvres around a low-flying Catalina, his Kittyhawk crashed into the sea, and he was killed.

Such was his memory, that forty years later, the Bryant and May matchbox company featured him on their *Redhead* matchbox covers. Many years after that, during the AFL's 2023 Anzac round, the Melbourne Demons wore a jumper which featured a Spitfire on the front and a tribute to those ex-players who had paid the supreme sacrifice, including Keith Truscott.

He was soon to become the squadron's "pin-up boy" and Australia's sixth highest scoring ace of the war. Indeed, he became such a household name in Australia that when he arrived home and led the Melbourne football team against Richmond, fans turned out in strength to welcome him. The Melbourne *Herald* reported:

> With the same smile that he wore when he last played football in Melbourne - in the 1940 grand final – Truscott responded to a rapturous greeting by 20,000 people when he appeared on the ground to play again for Melbourne. Arriving only ten minutes before the game, he had to push his way through the excited throng to get to the dressing room. Telegraph boys with messages of welcome and goodwill hammered at the door, which was barred to all except Melbourne players and officials. A great roar signalled his entry to the field. He acknowledged it with a wave and in a moment was surrounded by the Richmond team.

The big shock of the day, however, was the loss of Wing Commander Douglas Bader, the renowned tin-legged ace. Bader was leading the Tangmere Wing which was also on the operation and was shot down and captured near St Omer. He was interned as a POW. Bader had gained fame after he lost his legs in a pre-war flying accident. Fitted with prosthetic limbs, he went on to become a twenty-plus ace between 1940 and 1941. He was responsible for the RAF's "Big Wing" concept which he developed during the Battle of Britain whereby large, massed formations of fighters would take on the enemy bomber streams. The concept was detested by many fighter pilots as it took so long to form up by which time the enemy had usually bombed and departed. Bader's story is well known and will not be elaborated on here further. Needless to say, after several escape attempts, he was moved to Colditz Castle where he spent the rest of the war.

Circus operations continued, often without sighting the enemy. Such was the case on 12 and 14 August. But on 16 August this changed with No. 452 Squadron shooting down eight for no loss - their new Spitfire Mk Vbs were put to very good use. The morning began with *Circus 74* escorting six Blenheim bombers attacking the shell factory at Marquise. Finucane saw the enemy fighters manoeuvring to attack elements of No. 485 (New Zealand) Squadron, so he climbed to height and immediately shot one down. He later recalled:

> As we were leaving the coast near Gravelines, eight to ten enemy aircraft attacked 485 Squadron. I climbed to cut off further attack and saw one ME109E half roll towards ground. I took a quick burst from 250 - 300 yards and half rolled after him. I quickly overhauled him and gave a two second burst from 100 yards. Bits flew off enemy aircraft and he went into a spin. I broke off combat at 5,000 feet. E/a was last seen by me at 3,000 feet still in spin and going down vertically. Black One (Pilot Officer Truscott) confirms this. He saw e/a at 1,000 feet, still spinning and considers it impossible to pull out. The e/a was too low for him to attack without his a/c hitting the water.

No other Australians were able to get into a suitable firing position during the mission.

Later that evening, *Circus 75* was directed against the Longuenesse aerodrome at St Omer. As the squadron crossed the Channel, about fifty of the enemy were seen as they came up and crossed the French coast. The Messerschmitts attacked and No. 452 Squadron quickly became embroiled in a hot battle. Truscott recalled his role in the engagement:

> I was 452 Squadron, Green One. The squadron was attacked on return from target St Omer, and I did a steep turn and chased a 109 to about 9,000 feet but he dived away and I was then in the company of about four other Spitfires. Seeing two more 109s coming from the rear, I did a tight steep turn and came round on the inside at the second 109. He did a steep turn which I followed and had no difficulty in turning inside him. He apparently did not know that I was behind him and was making an attack on another Spitfire in the section. He took no further evasive action than a steep turn. I fired about a six second burst of both cannon and MG and saw him roll over and go down with thick white smoke coming from his plane. I followed him down to 6,000 feet and saw him go down into field near a wood. F/L Finucane (leader) who I almost collided with

> in the attack saw the 109E go down also. I then climbed to about 10,000 feet and had a full beam attack on a 109 at about 300 yards but I did not see any effects at my fire. I claim this 109 as destroyed as it was definitely seen to go into the ground by both [our] squadron leader (F/L Finucane) and myself. After my steep turn in the first attack, I saw a Spitfire on my port side attack a 109 which went down with thick black smoke pouring from the engine, the Spitfire followed the 109 down. Although I did not know it at the time, it was Green Two (Sergeant Stuart) who made this attack.

Indeed, Sergeant Archie Stuart claimed the victory. The tally of seven for the night was no doubt satisfying especially since the squadron suffered no loss of its own.

The rest of the month was spent on similar *Circus* missions. On 19 August, during *Circus 81,* Finucane destroyed a Bf 109 and probably another. Truscott claimed a probable as well. But this time, the squadron was heavily engaged by *Jagdgeschwader 2* and lost Sergeant Dick Gazzard and Pilot Officer Bill Eccleton, with both killed. In addition, Flight Lieutenant Douglas had his aircraft badly shot up and Pilot Officer William Willis was superficially wounded, although both landed their damaged Spitfires safely back at base.

Attacking St Omer again during *Circus 87* on the evening of 26 August provoked the expected *Luftwaffe* reaction and No. 452 Squadron claimed a further two destroyed for no loss with Flight Lieutenant Douglas and Sergeant Stuart the victors.

A posed photo outside the dispersal hut at Kenley of No. 452 Squadron pilots watching skywards. Left to right are Sergeant Bruce Dunstan, Sergeant Ian Milne, Sergeant Keith Chisholm, Sergeant John Elphick (rear), Pilot Officer Donald Lewis, Sergeant Jack Donald (rear), Pilot Officer Truscott (pointing), Flight Lieutenant Clive Wawn and Sergeant Raife Cowan.

The next day, Finucane and Thorold-Smith each destroyed two Bf 109Es completing the effort for the month. This brought the squadron's tally to 24 Messerschmitts destroyed and three probables. Flying Officer Denys Lane-Walters, the unit historian, was proudly able to write:

> This was a record score for the month of August for the whole of Great Britain and an exceptionally fine achievement especially when it is remembered that the squadron has only been fully operational for a period of two months.

Lane-Walters would not remain long at No. 452 Squadron and was soon replaced by Pilot Officer John Pollard. Pollard, an Englishman, would become very familiar to all pilots. As the intelligence officer, it was his responsibility to write up the monthly unit history sheets and lodge combat reports as well as preparing intelligence summaries for wing headquarters. He would remain with the squadron until 1943.

Towards the end of August 1941, the *Luftwaffe* began to deploy a new fighter, the Focke-Wulf 190A, to its fighter squadrons on the Western Front. The FW 190 was far superior to the Spitfire Vb and its arrival came as quite a shock to the RAF. When the new German fighter appeared in the skies over the Channel, it was almost 30mph faster than the Spitfire Mk V and it could out climb and out roll the British machine. Of more concern, the FW 190 could "leave the Spitfire with ease". At least the Spitfire could turn more tightly, a handy characteristic in a dog fight.

Nevertheless, No. 452 Squadron continued its success against the enemy during the high summer of September and into October. The risks were high, even for non-combat sorties. On 7 September, Sergeant James Hanigan and Sergeant Ken Williams collided in mid-air and crashed near Hooley, Surrey, during practice manoeuvres. Both were killed. They were training in the local area and were making dummy attacks on each other when they got too close with the sad, but inevitable result.

The German Focke-Wulf 190 fighter which proved a serious adversary for No. 452 Squadron when it was deployed on the Western Front from August 1941.

Meanwhile, on 2 September's warm summer day, No. 452 Squadron was kept busy. Two convoy patrol sorties and a "scramble" to intercept an unidentified aircraft (which was eventually identified as friendly), were followed late in the forenoon by a *Roadstead* attack on shipping near Ostend. A *Roadstead* was a bomber escort operation to attack ships at sea or in harbour. Three Blenheims and a few Hurricanes comprised the striking force, escorted by a wing consisting of Nos. 242, 452 and 485 Squadrons. Ship and shore anti-aircraft defences were extremely active and on the approach to the target, one of the Blenheims from No. 139 Squadron was shot down. The other bombers, however, claimed the sinking of a 5,000-ton vessel and a small anti-aircraft *flakship*. As the formation withdrew Flying Officer Bill Willis saw one Blenheim returning at sea-level without escort. Accordingly, he and Sergeant Arch Stuart broke away to protect it as its slipstream was churning up the water, leaving a long white trail which was bound to attract enemy fighters. Within two minutes a pair of Messerschmitts arrived but both were intercepted and shot into the sea, one by Willis and the other by Stuart. The Spitfires then returned independently having lost sight of the Blenheim.

Later in the month, the squadron recorded three days of what must only be described as aerial mayhem. On 18 September, they lost Sergeant Fred Manning and Flying Officer Bill Willis with both killed, while Sergeants Ken Try and Arch Stuart became POWs. Tasked on *Circus 99* to escort twelve Blenheims attacking Rouen, the flight was attacked by several Bf 109s from 7/JG26 which claimed six aircraft shot down. In return, No. 452 Squadron claimed four destroyed, one probable and two damaged.

The planned sequence of events had turned into a shambles when over Beachy Head, No. 452 found a squadron of Hurricanes in company with the forward box of six Blenheims forcing No. 452 to take on the rear six. But the slower Hurricanes pushed the Spitfires back into a line astern formation, which was not tactically sound. Just before the target, No. 452's B Flight was forced up over the escort wing as the bombers turned. Meanwhile, A Flight found it had to move across to cover B Flight and that allowed the Messerschmitts to take their turn with diving attacks. The airspace then degenerated into a series of scrappy dogfights in which both sides made claims.

Sergeant Bill Willis who was shot down and killed over France on 18 September 1941.

Sergeant Arch Stuart was another pilot shot down over France on 18 September 1941. He survived to became a POW. (Pix magazine)

Two days later, Warrant Officer Ian Milne also became a POW, but No. 452 Squadron claimed seven destroyed and three probably destroyed in a major air engagement involving up to 50 Bf 109s. The same could be said for the battle a day later. On 21 September, the final action for the month, five 109s were claimed and two damaged for no loss.

In October, the No. 452 Squadron hierarchy changed as the RAF authorities realised the unit did not need further supervision. On 3 October, Douglas was posted to command No. 403 (RCAF) Squadron with his place as B Flight commander taken by Truscott.

Meanwhile, Bungey's training programme and administrative leadership had given all squadron members confidence at a time when RAF Fighter Command was seizing the initiative. Finucane too imbibed a sense of aerial superiority in squadron members with his practical and up-front leadership in the air. Finucane's philosophy was simple:

> The tactical side of the game is quite fascinating. You get to learn, for instance, how to fly so that all the time you have a view behind you as well as in front. The first necessity in combat is to see the other chap before he sees you, or at least before he gets the tactical advantage of you. The second is to hit him when you fire. You mightn't have a second chance.

The nervous wait for pilots at readiness waiting for mission orders. Left to right are Sergeant Keith Chisholm, Sergeant Ian "Morch" Milne and "Paddy" Finucane. Both Chisholm and Milne would become POWs.

As October progressed, there were further sorties across the Channel with three producing results. On 2 and 12 October three of the enemy were claimed but on 12 October Sergeant Chisholm failed to return. The squadron was tasked to escort 23 Blenheims heading for the Boulogne docks. A series of dogfights ensued when Truscott damaged two and Finucane shot down one, but Chisholm in Spitfire Mk Vb W3520 was shot down and parachuted into the sea near Berk-sur-Mer. He was picked up by a German launch and after several temporary confinements, was sent to Lamsdorf POW camp. Through subterfuge, he escaped three times and avoided the Gestapo. By May 1944 he was in Paris fighting with the Resistance. For more than two years and by shear tenacity he remained at large. Finally, he returned to England on 30 August 1944. For his exploits in Paris, he was awarded the Military Cross.

Then came Monday 13 October. That day, the Kenley Wing was tasked on *Circus 108A* to provide escort cover for four No. 139 Squadron Blenheims headed for Arques where a ship lift was underway. *Circus 108B* was sent to Mazingarbe and was tasked at the same time, but for *108B*, No. 452 Squadron was not involved. The *Luftwaffe* was there in numbers to meet both *Circus* raids and in the action that ensued, No. 452 Squadron claimed six destroyed, one probable and three damaged. Sergeant Eric Schrader's claim of a probable was later upgraded. But the *Luftwaffe* also claimed its share. From their records, *JG 2* claimed thirteen Spitfires and *JG 26* a further seven. A Blenheim was also shot down. In reality, only twelve Spitfires were lost the whole day with two others damaged. One of those Spitfires lost, AB852, was flown by Sergeant Edgar Jackson who was killed and another by Sergeant John Elphick in AD310 who baled out over the Channel. Fortunately, he was rescued after an hour floating in his dinghy and he was back flying within the week.

No. 452 Squadron's general report of *Circus 108A* to Fighter Command illustrates the action which began just after 1300, five miles south of Gravelines. It is repeated here in total:

> 452 Squadron formed the lowest layer on the port side of escort to four Blenheims, whose target was Arques ship-lift. The squadron took off [from] Kenley 1240, joining up with 485 and 602 Squadrons and set course for Manston. They rendezvoused with four Blenheims 1302 hours and crossed the French Coast near Gravelines. Intense heavy Flak was encountered going in fairly accurate in height and direction. Halfway between the coast and St Omer, the squadron were ordered to engage about 15 Me 109s below. The squadron dived down and fierce attacks were delivered at 8,000/7,000 ft.
>
> F/Lt Truscott, taking advantage of his extra height, caught up with the leading Me 109F. He opened fire astern at 150 yds, giving a three second burst closing to 100 yds. The Me 109F went down smoking very badly. He followed to 3,000 ft and saw the E/A crash in a wood NW of St Omer. Other pilots confirmed (Destroyed).
>
> In this same attack P/O Thorold-Smith (Yellow One) overshot but turned sharp to port and found himself in excellent position to attack one of the 4/5 Me109Fs. He fired at 50 ft with MG [machine guns]. The E/A was extensively damaged, and the pilot baled out. P/O Thorold-Smith narrowly avoided collision with the Me 109F (Destroyed). Immediately afterwards, he observed a Me 109F on Yellow Two's tail behind him. He climbed hard

Squadron Leader Ray "Throttle" Thorold-Smith at his DFC investiture with his fiancé and Corporal Charles Marshall, also of No. 452 Squadron.

to starboard and got a quarter attack one second burst of MG at 100 yds. The E/A was damaged, a piece of port wing flew off and it dived downwards (Damaged). P/O Thorold-Smith observed Yellow Two's a/c (Sgt Elphick) smoking slightly. Sgt Elphick endeavoured to get home, but his engine seized when he was near the coast and eventually baled out 10/12 miles west of Wimereux. He was rescued after an hour in his dinghy.

Sgt Emery also joined in the attack on the 109s who were attempting penetration to the bombers. He got on the tail of one and started firing from 300 yds. The E/A suffered damage to fuselage, cockpit and wing roots. Large pieces few off when Sgt Emery was only 30 yds away. The Me 109 made a last attempt to pull out of the dive into which he had gone. However, he failed and went down in a vertical dive for 4/5,000 ft (not more than 2,000 ft from the ground) with smoke pouring. P/O Sly saw this a/c with clouds of smoke suggesting burning. He, like Sgt Emery, was unable to observe longer owing to the presence of more Me 109s. They turned to rejoin the squadron.

F/Lt Finucane (Red One) and Sgt Schrader (Red Two) were unable to position themselves for an attack in the first dive. They turned climbing and four Me 109s dived down ahead of the section from the port side. F/Lt Finucane took the leader of the second pair of the four. He gave a three second burst at 200 yds closing in to 100 yds and the Me 109F went down in pieces and flames.

Pilot Officers John Elphick and Robert Armstrong. Elphick abandoned his aircraft over the Channel on 13 October 1941 and was later rescued while Armstrong would crash in Australia in 1942.

Sgt Schrader took the remaining Me 109F and gave a long burst of about 4 sec closing to 50 yds. Strikes on the tail unit and the underside of the fuselage were observed to result in pieces flying off. The E/A turned over on its back with clouds of smoke pouring from it. It was last seen by Sgt Schrader at 8,000 ft. F/Lt Truscott who was getting back to rejoin the remainder of the squadron saw a pilot coming down in his parachute at 5,000 ft. (Probable)

F/Lt Finucane meanwhile had been attacked by 2 Me 109Fs. He fired at one, giving a one sec burst causing pieces to fly off the tail plane and fuselage. It took violent evasive action and F/Lt Finucane left it because he had heard Sgt Elphick was hit. (Damaged)

Near the coast he ran into a Me 109F which did a climbing turn to port, exposing its belly to F/Lt Finucane's fire. He got a long burst closing in from 180 yds and the E/A crashed into the sea. (Destroyed) During F/Lt Finucane's second combat, Sgt Emery was hit by empty shell cases from his fire. He engaged a second Me 109F and saw effective strikes on the fuselage and one cannon shell burst just behind the cockpit. (Damaged)

P/O Lewis had several skirmishes, and his No. Two (Sgt Jackson) is missing over France.

Eight of the Squadron circled Sgt Elphick until they were ordered to return to base owing to shortage of ammunition. P/O Coker's a/c engine was hit and eventually went out, and he made a fine belly landing at Kenley.

Six Spitfires landed at Kenley, three at Lympne and one at Hawkinge. All eventually returned to Kenley that evening. During *Circus 108A*, the *Luftwaffe* lost 22 aircraft plus three probables and eighteen damaged. The RAF lost twelve Spitfires, one of whom was Jackson's and another was Elphick's.

After the excitement of 13 October, just over a week later on 21 October, No. 452 Squadron packed up and moved six miles further south to RAF Station Redhill, a satellite aerodrome of Kenley. By late 1941, Redhill had seen its share of squadrons and action. Established in the early 1930s as an alternate civil landing ground to London's Croydon commercial airport, it became home to a service flying training school from 1937. At the start of the Battle of Britain, this school was relocated, and the station became fully operational for the defence

No. 452 Squadron pilots awaiting a scramble in the dispersal hut at RAF Station Redhill, where they were based from October 1941. Rear seated left to right are Sergeants RH Bevan, WJ Wilkinson, JR Ross and JM Morrison. Seated around the table left to right are Sergeant DFK Downes, Sergeant KD Bassett, Flight Lieutenant KW Truscott, Pilot Officer DF Evans, Flight Sergeant CJH Baldwin (RAF ground staff) and Pilot Officer RH Whillans.

of London and the south. By mid-1941, Redhill was a fully-fledged Spitfire base and No. 452 arrived to join No. 485 Squadron, as well as several other British Spitfire units.

November appeared to be a repeat of the two previous months with the usual *Circus* operations expected. The first, on 6 November, was a squadron escort of army cooperation Tomahawks over Cap Gris Nez. It was soon intercepted by a section of JG 26 and while the squadron claimed three aircraft destroyed and three damaged, the Germans shot down Sergeants Bernard Geissmann and Eric Schrader, both of whom were killed. One of the German aircraft destroyed was a FW 190, the squadron's first.

Thorold-Smith was the victor over the FW 190 and later recalled the combat. He looked over his shoulder and saw the FW 190 lining up to attack. Continuing the turn to port, he pulled up firmly into a very steep climbing turn, throttled back and flicked his Spitfire into a vertical turn in the opposite direction, so that the enemy was then below, slightly ahead and crossing him from port to starboard. He did a diving starboard beam-to-quarter attack, opening fire in short bursts at 300 yards, and closing to 200 yards and saw pieces begin to fly from the FW 190's engine cowling and tail. It went into a vertical dive leaving a black smoke trail. Thorold-Smith followed it down to 5,000 feet in case this was a breakaway feint but saw the aircraft crash into the sea. He also claimed a BF 109 in the same engagement.

The next day, operations changed to *Ramrods*. These had begun in late August, but *Ramrod 6A* would be No. 452 Squadron's first. A *Ramrod* was similar to a *Circus*, but its principal objective was the destruction of the target. On 7 November the target was the alcohol distillery at Beauchamps. While the bombing was to good effect, no claims were made by either side.

The hunting season of 1941 ended quietly with *Circus 110* on 8 November – it would be the last *Circus* No. 452 Squadron would fly. Because of the increasing losses the RAF was suffering during *Circus* operations, Fighter Command suspended them. The new FW 190s combined with a better early warning system now gave the Germans the upper hand. Truscott claimed two Bf 109Fs on 8 November but ended up in the water himself together with Sergeant Bruce Dunstan. Of the pair, Truscott was the luckiest. Baling out at 3,000 feet, his parachute harness became stuck on the seat. He finally freed himself from the cockpit at only 400 feet and his parachute opened just a few feet above the water. Both pilots were picked up by air-sea rescue launch and returned to the squadron. The remainder of November fizzled out with the start of the winter weather and there was little action.

Between August and November 1941, No. 452 Squadron held the Fighter Command record as the top-scoring unit, in part due to the fighting quality of its pilots, particularly Finucane, Truscott and Thorold-Smith.

In the early days of May 1941, the composition of the squadron's airman ranks was mostly made up of RAF members, but by the end of November it could boast 71 RAF and 81 RAAF airmen on strength. As such, it drew attention from the Australian politicians and media, all keen to show how the Australians were contributing to the war effort. No. 452 Squadron had quickly developed an enviable record and had become truly "Australian" in nature. Public interest back home was increasing, particularly with well-known VFL football star "Bluey" Truscott's rising status as an ace.

Consequently, the unit had its fair share of VIP visitors. The first was the Australian High Commissioner to Britain, Stanley Bruce with Air Marshal Sholto Douglas, Air Officer Commanding-in-Chief Fighter Command who visited on 21 August. Later, Former Australian Prime Minister Sir Earle Page, Air Vice-Marshal Richard Williams, Air Commodore Frank McNamara, VC, and the Kenley Station Commander, RAF Group Captain Cecil Bouchier, were among the visitors. Many photos of these VIP visitors are held by the Australian War Memorial.

No. 452 Squadron's opponents also had their aces. In November, the *Luftwaffe* in the west underwent a change of commander. *Oberst* Adolf Galland, an ace later credited with 104 victories (50 of which were Spitfires), was appointed *General der Jagdflieger* (inspector of fighters) after the accidental death of *Oberst* Werner Mölders in an aircraft crash. Galland had previously been commander of the fighter wing *Jagdgeswader* 26, and handed command over to Major Gerhard Schöpfel, another ace who would end his career with a score of 55. It would be the Bf 109s of JG 26 that No. 452 Squadron would regularly engage in 1941 and 1942.

On 8 December No. 452 Squadron encountered a flight of FW 190s off Dungeness and while a melee ensued, no enemy were shot down. However, Sergeant John Emery was shot down and killed. Emery was flying as Black One in Spitfire AB966 and was last seen chasing a FW 190 and firing at it towards the French coast, but two other FW 190s were seen chasing him. Despite the one-sided engagement, the aerial dogfight gave good intelligence on the performance of the FW 190 in combat. From pilot reports, they provided a quick assessment of the aircraft:

> The FW 190s undersurfaces were duck egg blue upon which black crosses showed distinctly, wing tips were yellow, tail fin light coloured like Me 109s. Noses resembling Harvards. E/A appeared faster in dive and climb than our A/C but S/Ldr Bungey and other pilots consider that it is not hard to turn inside them.

However, operations in December slowed considerably with the onset of the winter weather and much reduced hours of daylight. Protection of minelaying vessels, convoy escorts and a single uneventful sweep accounted for most of the eighty-two operational hours - a considerable drop from the 137 hours flown the month prior. This too was a fraction of the 445 hours flown in August.

Officially between June and December 1941 the RAF lost 411 fighters over the Channel and claimed 731 enemy fighters destroyed. In fact, the Germans recorded only 103 fighters lost.

A party of VIPs visits No. 452 Squadron. Left to right are Major Patrick Coleman (Assistant Secretary, Department of Air), Sir Earle Page (Australian Ministerial Envoy to the UK), Group Captain Cecil Bouchier, RAF (Officer Commanding Kenley Sector), Flight Lieutenant Keith Truscott and Air Commodore Frank McNamara, VC (Australian High Commission).

The overinflated claims of the RAF squadrons may have been allowed for morale purposes, but it is impossible now to say how many of No. 452 Squadron's claims were valid and how many were honestly claimed but were in error. Despite these inconsistencies, in 1941, No. 452 Squadron claimed 57 enemy aircraft destroyed, five probably destroyed and fifteen damaged. For the four months between August and November 1941, No. 452 Squadron held the record in the RAF for the most enemy aircraft destroyed across Fighter Command. A tally of these claims and squadron losses can be found in Appendix 4.

In the six months since No. 452 Squadron became operational, they too had lost their share of pilots – nineteen in all. Eleven had been killed in action and two were lost due to a mid-air collision. Six became POWs of whom Roberts and Chisholm escaped. The other prisoners waited out the war. The new year would bring a change of pace and several moves, but for the first three months, the aerial fighting continued.

A Pix magazine photograph of No. 452 Squadron pilots who were often featured in Australian media. Left to right are Sergeant Robert Anderson, Sergeant Ed Jackson, Sergeant Raife Cowan, Sergeant Eric Schrader, Sergeant Jack Donald, Flight Lieutenant "Paddy" Finucane (with pipe), Flying Officer Denys Lane-Walters and Sergeant Jack Emery. (Pix magazine)

The Uxbridge fighter control room from where No. 11 Group directed fighter operations, including that of No. 452 Squadron between July 1941 and March 1942.

No. 452 Squadron pilots in a group photo at Kenley in February 1942. Rear left to right are Flight Lieutenant RHC Sly, Sergeant DFK Downes, Pilot Officer CP Lloyd, Flight Lieutenant CN Wawn and Sergeant WJ Wilkinson. Centre left to right are Pilot Officer JRH Elphick, Flight Sergeant J James (RAF), Pilot Officer ME Sheldon, Pilot Officer DF Evans, Sergeant JM Morrison, Flight Sergeant FP Shelly, Pilot Officer RH Whillans, Sergeant WJ Smith, Pilot Officer RR Williams, Flight Lieutenant WA Seldon, Squadron Leader KW Truscott and Wing Commander JAG Gordon (RAF, Kenley Station Commander). Front left to right are Sergeant RJ Cowan, Pilot Officer R Armstrong and Pilot Officer JR Ross. Of those shown here, six did not survive the war including Wing Commander Gordon.

CHAPTER 3

JANUARY TO JUNE 1942 · AN INCREASING PACE

The New Year heralded a change of pace for No. 452 Squadron, with bad winter weather also playing a major role in slowing things down. *Circus* operations had virtually ceased due to increasing RAF fighter losses, particularly as a result of the new Focke-Wulf 190 fighters introduced into service by the *Luftwaffe* at the end of the previous year. Pressed on the Russian Front since mid-1941, the *Luftwaffe* had taken stock of its fighter force facing the Allies in the west and made changes. *Luftflotte* 3 (Air Fleet 3) defended France, Belgium and the Netherlands with two fighter groups known as *Jagdgeschwader*. By December 1941, these were *Jagdgeschwader* 2 (JG2 "Richthofen") flying Messerschmidtt Bf 109s and *Jagdgeschwader* 26 (JG26 "Schlageter") flying Bf 109s and FW 190s. Each *Jagdgeschwader* comprised three *Gruppen* (wings) of three *Staffel* (squadrons). No. 452 Squadron would face all of them.

Meanwhile, *Circuses* were replaced by operations called *Ramrods*, *Rhubarbs* and *Rodeos*. As previously mentioned, *Ramrods* were similar to *Circuses* but the prime objective was the destruction of the target. *Rhubarbs* were small-scale raids into enemy territory to attack enemy aircraft or ground targets, while *Rodeos* were fighter sweeps without bombers. These raids were by no means small as between March and June 1942, no less than 22,000 fighter and 700 light bomber sorties crossed the Channel. For the RAF, 303 fighters and eleven bombers did not return.

The first *Rhubarb* strafing target in the new year for No. 452 Squadron was the alcohol distillery at Colleville. This was attacked on 3 January by Thorold-Smith and Williams. They both scored hits on storage tanks but little else was reported. The next day, Pilot Officer Al Baker spun into the sea during a cannon test just off Beachy Head. Baker was killed when he hit the water, with the cause unknown. This was followed by another accident on 22 January during a minesweeper protection sortie. Flying Officer Don Lewis's Spitfire engine seized, and he too crashed into the sea. His orbiting colleagues could see a floating, lifeless body still attached to a parachute, but by the time the air-sea rescue craft arrived, his body was nowhere to be seen. Unfortunately, this outcome was all too familiar, with the freezing cold waters of the North Sea quickly taking a life. Adding to the problem, a Spitfire sank rapidly leaving little time for even a conscious pilot to escape.

Don Lewis was born in Hobart in February 1922. He listed his occupation as a survey draughtsman. After pilot training at Narromine, he was given further training at Camp Borden, Ontario, Canada, before being posted to No. 57 Operational Training Unit at Hawarden in Wales. There, after ten months of training in three countries, Pilot Officer Lewis completed his Spitfire conversion. Then it was onto No. 452 Squadron in May 1941 where he would spend eight months and seventeen days before his luck ran out, ten days short of his twentieth birthday. Like Al Baker, he is now remembered on the Runnymede Memorial and on Panel 105 at the Australian War Memorial.

With the new year, "Paddy" Finucane returned, a move that clearly lifted spirits. A break from operations to recover from his ankle injury had clearly done him good, as this letter to his mother in late January 1942 shows:

> I have arrived at Kenley and have settled down to the old way of living again. It is a grand feeling to be back with the chaps from the squadron. The Australians have [left] Redhill and now work from Kenley. I was posted back to the squadron yesterday as supernumerary. This means I can fly with them again. I am going to stay with them until a new job turns up for me.

Fight Lieutenant Clive "Bardie" Wawn who became B Flight leader in early 1942.

His stay was short as his posting to command sister RAF unit, No. 602 (City of Glasgow) Squadron, arrived just days afterwards and he left on 25 January.

The new year brought a change of airfield with No. 452 Squadron moving back to Kenley on 14 January 1942. Then, two weeks later, there were further changes. The commanding officer, Squadron Leader Bob Bungey, was given acting wing commander rank and was posted to RAF Shoreham, so Truscott was promoted and became the squadron commanding officer. As Finucane was promoted and went over to No. 602 Squadron also at Redhill, Flight Lieutenant Frank Coker became the commander of B Flight. Ray Thorold-Smith remained in charge of A Flight. Even Coker wouldn't last long as within a month he had been posted to No. 41 Squadron. He was replaced by Fight Lieutenant Clive "Bardie" Wawn.

Bob Bungey too, had earned a rest. He had been in almost continuous flying operations since the start of the war. Indeed, he was an Australian who had joined the RAF before the war on a short service commission. He had flown during the Battle of France in 1940 in the Advanced Air Striking Force with No. 226 Squadron, operating the obsolescent Fairey Battle light bomber. He then transferred to Fighter Command and flew Hurricanes during the Battle of Britain with No. 145 Squadron before taking up command of No. 452 Squadron two months after it had formed. He would mould the squadron into one of the RAF's finest and his "gallant and effective leadership" is one reason he was awarded the DFC in October 1941. Today, Bob Bungey is considered one of Australia's "Few".

Meanwhile, the long-awaited breakout of the battlecruisers *Scharnhorst* and *Gneisenau* and the heavy cruiser *Prince Eugen* was reported at 1040 on 12 February by Group Captain Victor Beamish, commander of the Kenley Wing and his wing leader, Wing Commander Bob Boyd, while on a reconnaissance over the Channel. The date would go down as a missed opportunity, but at least No. 452 Squadron played an important role. Eleven Spitfires were squadron scrambled at 1320 and proceeded to RAF Manston to rendezvous with forty Beauforts, but these did not appear. So, the squadron found the German flotilla and as there was no German fighter cover to contend with, attacked a German destroyer with cannon fire. The flak was intense but significant damage was done without loss. However, the main force of RAF and Royal Navy aircraft missed the three German major warships, which led to significant repercussions and soul searching by RAF higher command.

Aerial combat restarted in February. On 13 February the hapless victim was a Heinkel 114 floatplane which was brassed up by most of the squadron. Led by Beamish, Truscott and others shared in the complete destruction of the aircraft. This was No. 452 Squadron's 58th claim and the first for 1942. Then followed a patrol scramble two days later with Flight Sergeant Paul "Scruffy" Makin claiming a Bf 109 damaged. However, the squadron lost another pilot, Sergeant Frank Harper, who was shot down by *Leutnant* Hans-Georg Dippel of 2/JG26. The pair had been chasing the Messerschmitts which separated and although Makin got in a long burst which left the German aircraft smoking, it got away in a dive towards the French coast. Sadly, Harper was not so fortunate.

The hunting was better the following month with three *Circus* engagements on the 9, 13 and 14 March but losses also mounted. An uneventful *Circus 112B* on 9 March against the Abbeville marshalling yards saw No. 452 Squadron providing escort cover with the other two squadrons in the wing. Although the RAF lost three pilots during this mission, none were from No. 452 Squadron. This was followed by *Circus 113* on the same day which brought the German fighters up and these were engaged by the Kenley Wing. The target was the Mazengarbe fuel depot to be bombed by six Bostons of No. 107 Squadron, all of which bombed and returned safely. The wing provided escort, and the gaggle (truly a "circus") were supported by nine other fighter squadrons. Truscott and Wawn each claimed a Bf 109F destroyed, with Wawn claiming one probable and Truscott claiming one damaged. Sergeant Malcolm Hamilton was shot down and killed.

Hamilton was another typical No. 452 Squadron pilot. Born in Brisbane on 5 May 1917 and educated at Brisbane Grammar School, he joined the RAAF in November 1940. After pilot training, he was sent to England where he joined No. 52 Operational Training Unit at Aston Down in the Cotswolds to convert to Spitfires. Following conversion, he joined No. 452 Squadron in early November 1941 as a sergeant pilot. He survived four months and five days in the squadron before his death at age twenty-four.

Spitfire Vc and Focke-Wulf 190 Technical Data

	Spitfire Vc	FW 190A-3
Power Plant	1,415 hp Merlin 46 V-12-cylinder liquid-cooled piston engine	1,700 hp BMW 801 14-cylinder air-cooled radial engine
Dimensions		
Span	36 feet 10 inches	34 feet 5.5 inches
Length	29 feet 11 inches	28 feet 10.5 inches
Height	11 feet 5 inches	12 feet 11.5 inches
Weights		
Empty	5,065 lbs	6,393 lbs
Loaded	6,622 lbs	7,110 lbs
Max	6,700 lbs	8,770 lbs
Performance		
Max Speed	371 mph	400 mph (approx.)
Cruising Speed	226 mph	278 mph
Rate of Climb	3,250 ft/min	2,830 ft/min
Service Ceiling	37,000 ft	32,800 ft
Normal Range	470 miles	497 miles
Max Range	1,170 miles	582 miles
Armament	2 x 20mm cannon & 4 x 0.303-inch MG	4 x 20mm cannon & 2 x 7.9mm MG

Spitfire Mk Vb AB198 (UD-R) as flown by Flight Sergeant Ron Bevan and named Snifter II.

Circus 114 was flown four days later against the Hazebrouck marshalling yards and again, the wing put in a maximum effort. Pilot Officer Ray Sly destroyed a FW 190 and the squadron had no loss. Ray Sly would not make ace status of five confirmed kills, but he came close. He was officially credited with two destroyed and one probable plus two shared probables and one shared damaged. Unfortunately, he too would not see out the war. After a stint in sister RAAF unit No. 457 Squadron, he was posted to Malta in May 1942 as a flight commander of No. 185 Squadron, RAF. While leading the second batch of Spitfires to be launched off the

Sergeant Ron Bevan beside his Spitfire Mk Vb AB198 (UD-R) named Snifter II in mid-1942. Snifter was a popular cartoon character at the time.

carrier USS *Wasp,* he landed at Hal Far aerodrome. Then on seeing enemy fighters attacking those behind him, he attempted to take-off downwind but hit the top of a dispersal bay. His Spitfire crashed and exploded. Sly was extricated from the wreckage alive but succumbed to his injuries that afternoon. He was just twenty years old.

Flying Officer Ray Sly who was killed in Malta in May 1942.

By coincidence, another Sly arrived at No. 452 Squadron in March 1942. He was Flying Officer Ted Sly, no relation to the previous, but one who would find his time on European operations very short. In his memoirs he recalled his stint:

> We commenced familiarisation flying on 8 March 1942. That involved getting used to the area plus the fact that we were now flying Spitfire Mk Vs. The Spitfire 5 was quite an up-market plane compared to the Mk I. The rule when you first joined the squadron, was that you got the worst

aircraft but even though we were flying clapped-out 5s, we commenced doing sweeps of the Channel. We were just starting to settle in when 452 Squadron was posted out of operations to the Isle of Man.

Sly was shortly posted to the Middle East with No. 450 Squadron, RAAF, but would rejoin No. 452 Squadron in 1945.

Flying Officer Ted Sly shown in the Northern Territory. Coincidently he joined No. 452 Squadron just after his namesake Flying Officer Ray Sly was posted elsewhere.

Consequently, 14 March 1942, would be No. 452 Squadron's last major action in the European war. To celebrate the award of a bar to his DFC, "Bluey" Truscott shot down a FW 190, his thirteenth confirmed kill. His wingman, Sergeant John Morrison claimed a Bf 109 "probable" but was almost a victim himself. He landed with half his tail shot away during the skirmish after a running fight all the way across the Channel. He left No. 452 Squadron in April 1942 on return to Australia but was killed in a flying accident while converting to Kittyhawks with No. 75 Squadron two months later. As well as Truscott, "Bardie" Wawn, Pat Tainton and Jack Elphick also departed for home to bolster the defences back there. Squadron Leader Thorold-Smith became the new commanding officer but given the cracking pace No. 452 Squadron had endured over the previous year, and the loss of the veterans, the RAF authorities decided it was time for a rest and to rebuild before becoming operational again. Hence No. 457 Squadron replaced No. 452 Squadron at Redhill.

Sergeant John Morrison looking rather pleased with himself after surviving having his tail shot away on 14 March 1942.

Then on 21 March 1942 No. 452 Squadron moved to RAF Station Andreas on the Isle of Man. Upon arrival, they were met by the RAF station commander, Wing Commander Ed Knowles, DFC. As No. 452 had replaced No. 275 Squadron, RAF, an air-sea rescue squadron flying Lysanders, the incoming squadron picked up the task of rather monotonous convoy patrols and searching for downed pilots. Such was the case on 27 March and although Blue,

White and Pink sections got airborne to search for a Spitfire that had gone down twenty miles to the southwest, only an oil slick was located, and nothing was found of the pilot or aircraft. Within two weeks, No. 452 Squadron would lose another pilot, Sergeant Ralph Waters, when his Spitfire BL384 crashed into the sea eight miles southwest of St Bee's Head when returning to Andreas after an exercise off the Cumberland coast. Despite a search, little was found and the cause of the crash remains unknown.

RAF Station Andreas is located on the northern tip of the Isle of Man just a few miles north of the town of Ramsey. The land was requisitioned by the RAF in 1940, and an airfield was constructed which was opened in 1941. It had three cross runways each of about 1,300 yards long. The station initially hosted an air-sea rescue squadron but in October of that year, No. 457 Squadron arrived before they were replaced by No. 452 Squadron in March 1942. No. 452 Squadron's stay at Andreas was short as they and No. 457 would leave for Australia in June. Andreas was about as far away from the action that a fighter squadron could get, but the RAF authorities had conceded the unit needed a rest. Nevertheless, new and untried pilots arrived to replace losses, and their inexperience would prove fatal for two and cause the loss of several Spitfires.

On 7 May, Pilot Officer Robert Whillans was forced to bale out of Spitfire AB171 due to engine failure and subsequent fire while attempting to force land at Jurby, about four miles from the squadron's airfield at Andreas. He would never have made it to Andreas and the aircraft crashed. Fortunately, Whillans was found safe and would go with the squadron to Australia. The next day, there was further tragedy. Sergeant Reg Goodhew and Pilot Officer William Ford collided during practice camera gun attacks. Goodhew was killed during a head-on attack which he was leading but was forbidden to do when the two aircraft collided. Ford baled out and returned to the squadron, but he would be killed while still with the squadron in February 1943 during a transit flight from Darwin to Wyndham.

***

On the other side of the world, the 1941-42 Japanese onslaught through the Netherlands East Indies seemed unstoppable, and after the bombing of Darwin in February, Australians feared the worst. Political pressure on Prime Minister Curtin began to grow to bring Australian forces home to protect the nation. Fear of Japanese invasion was palpable. The pressure grew even further after Justice Lowe of the Victorian Supreme Court presented his damming Royal Commission report into the Darwin bombing and how unprepared Australia's defences were. Calls quickly mounted to bring RAAF fighter squadrons back to defend the country. Given mounting pressure from the media and populace, the Curtin government dispatched the Attorney General and Minister for External Affairs, Herbert "Doc" Evatt to Washington and London in May 1942 to seek a greater share of war resources. In Washington, he received a sympathetic hearing and acquired a large aircraft order.

In London, Evatt met with Churchill to plead for Spitfires and an aircraft carrier for the Royal Australian Navy, and while the carrier was not forthcoming, an agreement was reached on 28 May 1942 to supply the Spitfires. Nos. 452 and 457 Squadrons would be withdrawn from

Prime Ministers Curtin and Churchill meet in April 1942. As a result of this meeting Churchill agreed to send Australia a wing of Spitfires to defend Darwin.

England to Australia, while No. 54 Squadron, RAF, would be included in the deal. These three became known as the Churchill Wing. The arrangement was that each squadron would receive sixteen aircraft as initial equipment and replacements would be provided at the rate of five aircraft per squadron per month. In an effort to confuse the Japanese, the Spitfires would be called *Capstans* (after a popular brand of cigarettes) in any correspondence or future press releases. However, it is highly doubtful this subterfuge deceived anybody least the Japanese because as soon as the *Capstans* began flying in Australia, their elliptical wing shape was a dead giveaway. The Australian public was only made aware they were Spitfires on 4 March 1943 after Raid 52 over Coomalie Creek in the Northern Territory.

Then as always, as far as combining RAF and RAAF squadrons, the devil would be in the detail. First, there was the matter of parity of pay and conditions which had been festering since late 1942. By mid-1943 119 RAAF airmen were serving in No. 54 Squadron vis-a-vis 134 RAF airmen in Nos. 452 and 457, where differences in pay between the RAF and RAAF was causing disgruntlement. The Air Board proposed that the British should pay their servicemen British salaries and any Australians serving in British squadrons be paid the British rate by the British, topped up by the RAAF to meet RAAF rates of pay. The Minister for Air Arthur Drakeford agreed, but it took over a year to finally resolve the dispute.

No. 54 Squadron, RAF

No. 54 Squadron, RAF, had a distinguished history. Formed in May 1916 for service in the UK and France, it disbanded in 1919 with the downsizing of the RAF postwar. It reformed in 1930 for the air defence of Britain and later fought in the Battle of Britain before, like No. 452 Squadron, it began sweeps and *Rhubarbs* over France in 1941 and 1942. It too flew a steady stream of Spitfire Mk 1s, IIs and Vs. By June 1942 it was at Castletown, Isle of Man, for a period of rest and recovery, so it made sense to combine it with the two Australian squadrons also there. It was commanded by Squadron Leader Eric Gibbs, RAF, who had joined the squadron from Coastal Command in April 1942, an unusual move. Gibbs took the squadron to Australia and remained its commanding officer until January 1944 when he was replaced by Squadron Leader Rob Newton, DFC. Gibbs became a five-claim ace while in Australia and was awarded a DFC for his leadership. The final two war years of 1944 and 1945 produced slim pickings for the squadron with ground attack sorties against airfields and transport escorts the norm. At war's end, the squadron moved to Melbourne for embarkation to UK and was officially disbanded there on 31 October 1945. Once home, it reformed as a fighter squadron for the air defence of the UK.

A No. 54 Squadron Spitfire flies over the Northern Territory.

No. 457 Squadron, RAAF

No. 457 Squadron was the other Australian squadron to be withdrawn from the UK. It was formed at Williamtown, NSW, like No. 452, albeit some months later in August 1941. The ground personnel embarked that month for the UK. Following the experience of No. 452 Squadron, Australian pilots and groundcrew already in the UK had established the squadron there in June 1941 on the Isle of Man flying Spitfire Mk Is and Mk IIs. Initial flights were convoy patrols with the occasional scramble to intercept unidentified aircraft. It was not until March 1942 when the squadron moved to Redhill in No. 11 Group that operations over France commenced. The squadron would only fly offensive sweeps and *Circuses* until May when it was retired to Kirton-in-Lindsey in preparation for the return to Australia. Once home, the squadron was based at Livingstone, a strip south of Darwin where it fought the war against Japanese raids and later up in the islands. No. 457 Squadron was officially disbanded at Labuan on 7 November 1945.

Spitfire ZP-Y of No. 457 Squadron decorated with a shark's mouth and named Grey Nurse, which became a nickname for the squadron. (Watson collection)

In preparation for the return to Australia, a No. 452 Squadron movement shuffle began. On 1 June they moved to RAF Station Atcham in Shropshire where all were granted seven days leave. On 12 June they moved to RAF Baginton in Warwickshire where "Doc" Evatt paid a visit followed by Air Vice-Marshal Frank McNamara, VC, the senior RAAF liaison officer in London who inspected the squadron and wished them farewell.

Niceties aside, on 18 June, the long journey home began. At 1900 the *Stirling Castle* weighed anchor from Princes Dock in Liverpool, only to wait in the Mersey for a convoy to which she would be part to form up. Finally, four days later she sailed with over 4,000 men (RAAF, RAF and army) aboard a ship designed to carry 1,700. Conditions were cramped and the men fractious. While at sea, several promotions came through for the officers and men so by the end of June, No. 452 Squadron could boast 22 officers and 208 other ranks. On 13 August, the ship arrived at Port Melbourne where the men found they had been posted to RAAF Station Richmond west of Sydney to prepare for a new theatre of operations.

A huge mushroom cloud billows over Darwin harbour as the merchant ship Neptuna, loaded with depth charges and anti-aircraft shells, blows up. This is one of the iconic images of the 19 February 1942 Japanese raid. Subsequently Darwin became the only location on the Australian mainland to be regularly attacked by Japanese aircraft. The Spitfires of No. 1 Fighter Wing would be responsible for the defence of the northern capital in 1943-44.

Curtiss P-40E fighters of the USAAF's 49th Fighter Group lined up for a 1942 media publicity photograph. This group defended the Darwin area during the 1942 dry season and would be replaced by No. 1 Fighter Group Spitfires the following year.

CHAPTER 4

JULY TO DECEMBER 1942 · A CHANGE OF SCENERY

When Japan concurrently attacked Allied bases across South-East Asia on 8 December 1941, the extent of the war changed markedly. So did the threat to Australia. By February 1942, the British and Australians had been defeated in Malaya and the Netherlands East Indies after relentless assault. The "impregnable" fortress of Singapore had fallen, and the Royal Navy's capital ships HMS *Prince of Wales* and *Repulse* lay at the bottom of the South China Sea. Despite Churchill's assurances of Britain coming to Australia's aid in time of need, reality proved otherwise. Australia was now alone as Britain was overextended in the Atlantic, North Africa and the Mediterranean. With the bombing of Darwin on 19 February 1942, the war became very real for all Australians.

To fight their Pacific War, the Japanese military had earlier developed two "air forces" – the Imperial Japanese Naval Air Service and the Imperial Japanese Army Air Service. Each had a separate role and fought separate wars. It was the IJN's role to fight the war in the Pacific, so the majority of the raids on Australia were led by naval forces of bombers, almost entirely Bettys escorted by formidable Zero fighters. The IJA only raided Darwin once, on 20 June 1943, with a mixed formation of Helen and Lily bombers escorted by Oscar fighters.

During April 1942, the Japanese reinforced their holdings in the Netherlands East Indies and New Guinea. This was done for two reasons. The first was to cement their foothold on captured territory to allow unfettered access to their resources, particularly oil and rare minerals from the Netherlands East Indies. The second reason was to isolate Australia (and New Zealand) from becoming an American launch pad for what would inevitably be a counter-offensive. In the latter case, they would not be successful.

After the fall of Singapore, Darwin would be the next city to face the Japanese onslaught but to meet the threat, the RAAF could only muster seventeen Hudson bombers in Nos. 2 and 13 Squadrons, and fourteen Wirraway trainers in No. 12 Squadron. There were no RAAF fighters present, and a token defence was briefly provided by ten American P-40E Warhawks in transit to Java.

Meanwhile, the Americans had withdrawn from the Philippines and Netherlands East Indies, with General MacArthur moving his headquarters first to Melbourne, then Brisbane while the US Army Air Force set up camp in three main locations: Darwin, Townsville and Brisbane. Darwin would be used for flying units as a vanguard against Japanese invasion; Townsville for repair, maintenance and supply; and Brisbane for deeper maintenance and aircraft assembly.

As far as Darwin was concerned, by mid-March the 49th Fighter Group, USAAF, had begun arriving, and this unit would defend the Darwin area throughout the 1942 dry season. Equipped with P-40E Warhawks, many of the pilots were initially inexperienced but they were well led and proved to be fast learners. However, it was a matter of national prestige that Australian fighter units should be defending Australian territory, and RAAF Kittyhawk squadrons replaced the 49th Fighter Group in the final months of 1942.

As discussed in the previous chapter, Prime Minister John Curtin had wanted Australian fighter squadrons to defend Darwin. As a result of Dr Evatt's visit to London in May 1942, Churchill promised Australia three Spitfire squadrons despite a deteriorating situation in the Middle East and North Africa. Nos. 452 and 457 Squadrons, along with No. 54 Squadron, RAF, were selected for redeployment. Forty-eight Spitfires would accompany them. It was a propaganda coup for Evatt and Curtin while ensuring the British would have a say in the postwar peace negotiations in the Pacific. At this time there was no announcement to the Australian public.

DAILY MIRROR, Saturday, July 18, 1942.

Daily Mirror

JULY 18

12,042 ONE PENNY

Registered at the G.P.O. as a Newspaper.

PADDY FINUCANE IS DEAD

ROME BREAK

"This is it, chaps" as plane hit sea: al avenged him

Paddy Finucane, the RAF's greatest "ace" fighter, has been killed in action.

RIGHT to the end Wing-Commander Brendan Paddy Finucane, D.S.O., triple D.F.C., and destroyer of at least thirty-two German aircraft, remained master of the Luftwaffe.

For it was no enemy plane that brought him down but a chance shot from a German machine-gun on the French beach near Pointe du Touquet.

The bullet hit his Spitfire's radiator. Paddy carried on to his target, turned for home and had to land on the sea.

The plane sank immediately, carrying Paddy with it.

Right to the end, too, Paddy continued to talk calmly over his radio.

His last message—probably as his engine stopped—was: "This is it chaps."

Swift Vengeance

Paddy's "Number Two," Pilot-Officer F. A. Aikman, wrought swift vengeance on the machine-gun post which fired the fatal shot.

"I guess my fire blew it pieces," he said.

The incident that en…

Pilot-Officer F. A. Aikman . . avenged his leader.

Wing-Commander Paddy Finucane, D.S.O., D.F.C. (double bar), who died unbeaten in the air.

1,000,000 Huns

war

The death of Finucane was front page news. This Dail Mirror article of Saturday 18 July 1942 describes him as "the RAF's greatest ace".

With the political deal done, a mixture of Australian and British personnel embarked in the liner *Stirling Castle* together with six Spitfires. With space limited, forty-two other Spitfires were loaded onto the freighter *Nigerstown.* The convoy departed Liverpool in June 1942, sailed around the Cape of Good Hope, crossed the Indian Ocean and arrived in Melbourne on 13 August 1942. Only six Spitfires were with them.

The main shipment of forty-two aircraft—enough to fully equip all three squadrons—was diverted to North Africa. On 21 June, Tobruk had fallen to Rommel's *Afrika Korps*, and three days later, the Dominion Office urgently requested that the Spitfires be offloaded at Freetown and flown to Egypt to replace losses. Curtin refused, arguing that Australia's need was no less pressing than that of the RAF in the Middle East. Despite his protests, Britain disregarded his argument, and the Spitfires were redirected leaving the RAAF to wait till October for forty-three Mk Vc reinforcements.

It was also during the voyage home that No. 452 Squadron members heard the sad news that "Paddy" Finucane, their one-time flight leader and mentor, had been shot down and killed. Not by the *Luftwaffe*, but by German shore guns that he was strafing on the French coast. "Bluey" Truscott took it the hardest. They had become close friends and as Finucane had declared the end of the war would not see the end of their friendship, perhaps he had a premonition. Truscott too would not see out the war, killed in a senseless aircraft accident off the Western Australian coast on 28 March 1943.

Westminster Cathedral was filled to standing room only for Finucane's official Requiem Mass celebrated by Monsignor Henry Beauchamp, Senior Catholic Chaplain to the RAF, in the presence of Cardinal Arthur Hinsley, the Archbishop. The congregation came from all walks of life, from senior military personnel and members of the public to adoring young women and schoolboys. Also in attendance were the senior USAAF chaplain in the UK representing the Americans, Air Vice-Marshal Frank McNamara, the RAAF's senor officer in London, and the Australian High Commissioner, Stanley Bruce.

Then came the many letters of sympathy to Finucane's mother. Here are a couple. The first was from Eve Moore-Johnson who wrote on 19 July 1942:

> Dear Mrs Finucane,
>
> My husband and I want to convey to you and your family our deepest sympathy on the loss of your dear son "Paddy". Such a grand gentleman, such an example to the entire world, we are proud to have shared "these times" with him...

The second was from Pat Hoare who wrote on 26 July 1942:

> Dear Mrs Finucane,
>
> I hope you will not mind me writing to you. My name is Pat. I am 12 years old. I feel I would like to express my sympathy with you on the loss of your dear son. I have been very interested in him and have followed his career. I keep a scrapbook and have a large number of paper cuttings of him. I have an uncle who is a bomber pilot and also another

uncle who is training for a fighter pilot and I hope I shall be as proud of them as you are of your son.

Best Wishes, Pat Hoare.

Despite the disappointment of having no Spitfires when No. 452 Squadron arrived back in Australia, a second batch of seventy-one aircraft arrived on 25 October 1942 with thirty-three more enroute – enough to equip the squadrons and provide for four months' combat replacements and training aircraft. The men had arrived in August where they were grouped into No. 1 Fighter Wing under the redoubtable Wing Commander Clive Caldwell. Training commenced at RAAF Station Richmond while they awaited their aircraft. At least the six original Spitfires had been flown to Richmond in early September which together with four Wirraways and two Ryan ST-Ms, allowed some training to commence. Nos. 452 and 457 Squadrons were each allocated two Spitfires, two Wirraways and a Ryan apiece.

Wing Commander Clive Robertson Caldwell

Wing Commander Clive Caldwell who led No. 1 Fighter Wing from November 1942. (Kenward collection)

Wing Commander Clive Robertson Caldwell, DSO, DFC*, Polish Cross of Valour, was well known across Australia for his "ace" status from his time in the Middle East where he had scored 20.5 kills against the Germans and Italians. He was to add another eight during his time back home. Caldwell had been born in Sydney and undertook flying lessons with the Royal Aero Club in 1938 before joining the RAAF in May 1940. After pilot training he was posted to the Middle East from May 1941 to May 1942. Returning to Australia in September 1942, with much acclaim he was appointed wing leader of No. 1 Fighter Wing at Darwin. With a move north into the islands in 1944, he took command of No. 80 Wing as a group captain. The wing moved to Morotai and then Tarakan to see out the war. During his time in Darwin and the islands, he regularly flew with the squadrons under his command. At war's end he was Australia's highest scoring ace.

Upon arrival in Australia, the Spitfires were called *Capstans* powered by *Marvel* engines and this is how they were referred to in the press. It is doubtful the subterfuge fooled anyone as even schoolboys could recognise the Spitfire's beautiful elliptical wing and the sound of a Merlin engine. The term was soon dropped once the Spitfires arrived in the Northern Territory and after the BBC announced the Spitfires had arrived well before the Australian government!

With the quarters at Richmond cramped for space, A Flight moved to Schofields on 16 September 1942 while B Flight camped across the airfield in what became known as "Little

A newly assembled Spitfire Mk Vc "Capstan" at Laverton, Victoria, soon after arriving in Australia in late 1942. This fighter was destined for No. 79 Squadron.

Pitt Street". The commanding officer and headquarters staff remained on the station. Practice flying commenced in October and it was not long before the squadron registered its first fatality in Australia. On 9 October Sergeant Michael Clifford had been tasked to conduct an oxygen climb to 35,000 feet when his aircraft unexpectedly crashed into the sea about a mile off Wamberal, Broken Bay. An oil slick was all that remained, but witnesses reported the aircraft dived straight into the water and there was no sign that the pilot had attempted to bale out. Clifford's body washed up the next day, and he was later buried in the Botany General Cemetery. Before his RAAF service, Clifford was an international rugby union representative for Australia and first grade cricketer, so was well known and his loss was keenly felt.

Indeed, October 1942 was to be a bad month. As well as Clifford, the squadron lost two more members in flying accidents. On 20 October Flight Lieutenant John Ross with Leading Aircraftman Walter Elliott as passenger took off in one of No. 452 Squadron's Ryan training aircraft (A50-21) for aerobatics around the Schofields base. The aircraft crashed about fifty yards outside the perimeter after it had pulled up steeply to avoid power lines. The aircraft stalled, the starboard wing dropped, and the Ryan entered into a spin before crashing into the ground. Both survived the crash, but Ross died without regaining consciousness in the station sick quarters while Elliott died the next day.

Given the increasing RAAF build up at Richmond, No. 452 Squadron relocated to Bankstown aerodrome from 11 November but were given no sign of a move north. Adding to the boredom, poor weather prevented much flying. By the end of the month the squadron could boast 22 aircraft and 24 pilots of whom seventeen were officers and seven were sergeants. More men were steadily being posted in, in readiness for a move to Darwin.

Apart from beat-ups and mock attacks between squadrons and "raids" on Camden aerodrome, there was little for the squadron to do. There were more accidents the following month. The first occurred on 13 December when Flying Officer Davidson Evans belly landed A58-65 (BS169) on Palm Beach after having engine problems. The unit historian recorded that:

> … a truck load of squadron personnel went up after tea to attempt to drag the machine from the water, but the tide overcame the efforts of all hands, and the aircraft was finally left in the small hours of the morning with the water just lapping the cockpit door.

Remarkably, the aircraft was recovered and eventually repaired and returned to service. Evans did likewise. He would end the war with no enemy claims but was rather lucky to have survived four Spitfire crashes during his time with the squadron.

Four days later, the next crash was more serious. Flying Officer Robert Armstrong had the starboard oleo of A58-91 (BS230) shear off on landing at Bankstown. The aircraft was catapulted forward and overturned trapping Armstrong in the cockpit. The emergency services had great difficulty extricating him. He was taken to Concord Hospital and placed on the "Dangerously ill" list. Happily, he later recovered although his flying days were over, so he became a fighter controller in 1944.

Then, just after Christmas leave, on 30 December, the squadron was finally ordered to move. Flying Officer Bob Whillans and the squadron doctor, Flight Lieutenant Bill Seldon, led the advance party with the remaining ground staff travelling by the ship MV *Maatsuyker* from 12 January. The air party comprising 23 Spitfires and a Lodestar transport aircraft all escorted by a Beaufighter left three days later. The flight took the Spitfires on a tour of the outback – via Camden to Wagga Wagga, thence to Mildura. Next, to Oodnadatta then Alice Springs followed by Tennant Creek, Newcastle Waters and Daly Waters and finally to an airstrip cut out of the bush called Batchelor. It took three days to cover the 2,200-mile trip. Thirteen days later, the sea party arrived. All set about preparing for operations.

Flying Officer Bob "Hal" Whillans who was co-leader of the No. 452 Squadron advance party that arrived in the Darwin area at the start of January 1943. (Kenward collection)

The other leader of the advance party, Bill Seldon, who was a medical officer. He is seen later in the war when he was a squadron leader. (Kenward collection)

Strauss airfield became home to No. 452 Squadron in January 1943. It was rudimentary with a single north-south runway and parking areas. Like most of the airfields in the Darwin area it was located alongside the main north-south highway for logistical purposes.

CHAPTER 5

JANUARY TO JUNE 1943 · FACING THE JAPANESE

Batchelor airfield is located forty-two miles south-southeast of Darwin down the Darwin to Katherine Road. Initially constructed in 1933 as a commuter airstrip, it was named after a South Australian politician, Lee Batchelor. The land was acquired by the RAAF in July 1941 for defence purposes. During the war, the field consisted of two cross runways, one sealed of about 1,700 yards long and the other oiled gravel of about 1,000 yards. Only the northwest–southeast runway remains today. For war operations, the RAAF installed a control tower and there was accommodation, messing and minor maintenance facilities as well as refuelling available. From July 1941 to September 1942, it was home to the Americans, but when they left, it became available for No. 452 Squadron. On 17 January 1943 the squadron arrived.

Within a fortnight, No. 452 Squadron was told to move, this time further north to Strauss which was half the distance to Darwin. Strauss was somewhat rudimentary having a single 1,700-yard gravel runway facing north-south. It was constructed in 1942 and had little in the way of facilities. Built by the Americans, it was named after USAAF Captain Allison W Strauss, whose P-40 Warhawk was shot down into Darwin harbour during a Japanese raid on 27 April 1942. Known also as Humpty Doo or 27-mile strip, the airfield became home to No. 452 Squadron from 1 February 1943 to May 1944.

While Strauss was remote, No. 452 Squadron members made the most of it. Tent accommodation was shared under the cover of heavy canvas which mostly kept out the

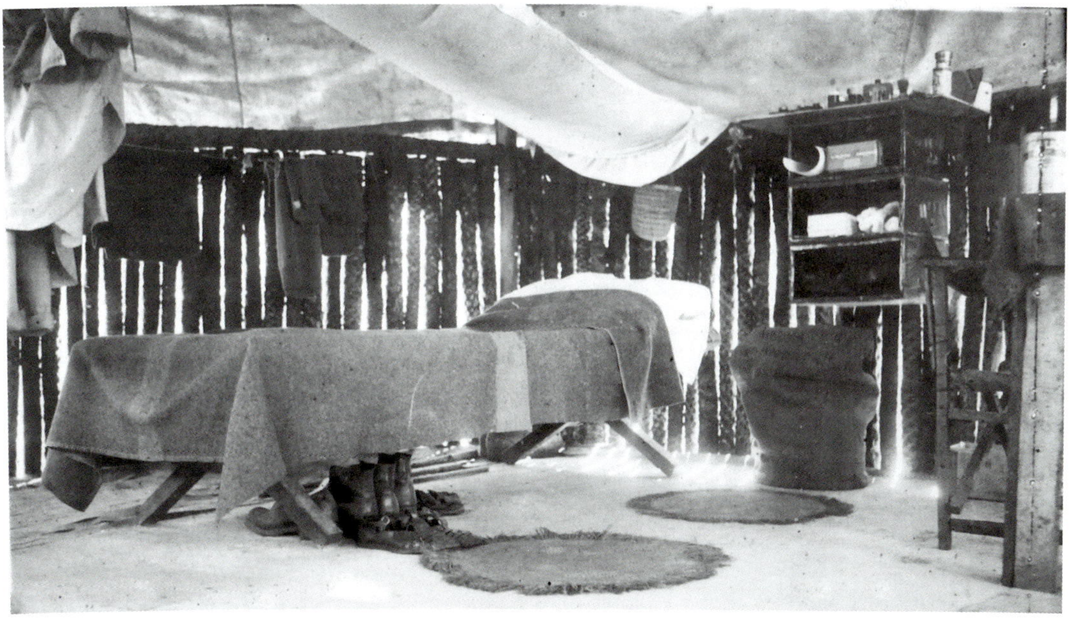

An example of the basic airman's accommodation at Strauss. (Kenward collection)

Members of the No. 452 Squadron medical section in front of the tented sick quarters at Strauss. (Kenward collection)

monsoonal rains. These were still prevalent in February although the end of the wet season was approaching, which would also see a resurgence of Japanese aerial activity. A camp stretcher with a blanket and pillow was provided for bedding. Scrounged boxes made suitable cupboards and empty drums, bedside tables. Other tents were erected for the medical section, stores and administrative use.

Just weeks after the Pacific War began, on 15 January 1942, the RAAF had formed Headquarters North-Western Area sited at the Darwin RAAF station. This headquarters would control RAAF units in the Northern Territory and others then based in the Netherlands East Indies. After the first bombing of Darwin on 19 February 1942, the RAAF needed to reorganise and do so quickly. As well as greatly improved defensive measures, the RAAF deployed No. 5 Fighter Sector for control and No. 31 Radar Station for early warning. Although Japanese raids continued throughout the year, by late 1942 the Darwin area had become important for aircraft staging, convoy protection and attacks on Japanese bases in Timor and the surrounding islands of the Netherlands East Indies.

The arrival of three fighter squadrons in the RAAF's North-Western Area was a huge boost to morale for those in the Northern Territory, and there were hopes that wave after wave of seemingly unstoppable Japanese attacks could finally be halted. The three squadrons (Nos. 452 and 457 Squadrons, RAAF, and No. 54 Squadron, RAF) comprised No. 1 Fighter Wing and operated under a wing headquarters. The wing was now commanded by Group Captain Alan "Wally" Walters with Wing Commander Clive Caldwell as his deputy and wing leader. By February 1943, No. 54 Squadron was at Darwin commanded by Squadron Leader Eric Gibbs, RAF, while No. 452 Squadron was at Strauss and No. 457 Squadron was at Livingstone. The two RAAF squadrons were commanded by Squadron Leaders Ray Thorold-Smith and Ken "Skeeter" James respectively. All were ready to make their mark.

Once "up north", No. 452 Squadron wasted no time in getting down to business. On 21 January 1943, they made their first operational patrol when White and Yellow Sections were scrambled to intercept a Bogey. A bogey is a suspected enemy aircraft, but in this case, it turned out to be a friendly Hudson returning to base.

Six days later, there was a tragic mid-air collision over Coomalie Creek. Sergeant Eric Hutchinson in A58-73 (BS184) clipped Sergeant Henry Stockley in A58-55 (BR549). Hutchinson's aircraft tail hit Stockley's port wing after which Hutchinson went straight down out of control with no chance of escape. Stockley, after entering a steep dive, managed to pull up and land with his wing badly buckled. Hutchinson's body and the smashed wreckage were found the next day. He was buried in the Adelaide River War Cemetery with full military honours, with Thorold-Smith and many his squadron mates in attendance. Of Hutchinson, the unit historian wrote:

> Eric Hutchinson, despite his short connection with the squadron, was most popular with his fellow pilots, and his quiet, unassuming manner earned him the respect of squadron personnel. He had a very fine physique and had excelled as a rugby football forward previous to joining the RAAF.

The move to Strauss went smoothly and the squadron was at readiness the next day, 2 February 1943, but they had to wait four days for action when an enemy reconnaissance plane was detected. Red Section (Flying Officer Adrian Goldsmith and Flight Sergeant Francis Turnbull) and Blue Section (Flight Lieutenant John Bisley, Sergeant Ross Stagg, Sergeant David Downes and Pilot Officer Fred Young) were scrambled to 35,000 feet, but no enemy was seen. The Japanese aircraft, however, did not survive as Flight Lieutenant Bob

Members of No. 1 Fighter Wing pose for a public relations shot in Darwin, with members of all three component squadrons present. Left to right are Sergeant Jim C Wellsman, Pilot Officer George B Jerries, Flying Officer Jerry A Tuckson, Flying Officer John D Fenegan, Flying Officer Davidson F Evans (RAAF), Flight Lieutenant Bob Foster (RAF), Pilot Officer David F Downes (RAAF), Pilot Officer Anthony K Brock, Pilot Officer Fred J Young (RAAF), Sergeant David M Wheeler, Flying Officer Ernest Weatherhead (No. 54 Squadron Adjutant), Flight Lieutenant Robin Norwood and Pilot Officer McIntyre. Aside from the four men indicated as members of the RAF or RAAF, the remaining men are all members of the RAF Volunteer Reserve. (Library and Archives NT, Weatherhead collection)

Foster of No. 54 Squadron shot it down off Melville Island. The aircraft was a Mitsubishi Ki-46-II Dinah flown by Lieutenants Kurasuke Setoguchi and Fumio Mori. Designed for high altitude reconnaissance, twin-engine Dinahs from the 70th *Dokuritsu Chutai* (DCS - Direct Command Squadron) flying from Lautem in East Timor would become regular visitors. The Dinah had a range of over 1,500 miles making it ideal for its role. A flexible rear-firing 7.7mm machine gun provided some measure of defence but in practice it was not always carried.

Before each Japanese raid, the 70th DCS would usually fly a reconnaissance to confirm the target and check the defences. A radio report back to base ensured the vital information was available for those in command of the next raid. On this occasion, Lieutenants Setoguchi and Mori were caught at 22,000 feet and Foster fired five bursts over twelve seconds. The last set the port engine ablaze which soon engulfed the fuselage. The Dinah went into the sea still burning. This was the first claim by Spitfires over northern Australia.

The interception been aided by Radio Direction Finding apparatus – basically a rudimentary radar. Located at Dripstone Caves just north of the Darwin town, No. 31 Radar Station provided early warning of Japanese raids in conjunction with spotters on Melville and Bathurst Islands. This service would prove invaluable in intercepting the regular Japanese raids during the 1943 dry season. The RDF consisted of a large square mesh antenna atop a mast and was camouflaged with tree branches to avoid detection. A control room similarly camouflaged under netting was located nearby. The original Dripstone Caves site was augmented by more substantial RDF towers at Lee Point and Cape Don.

Unfortunately, accidents continued with many fatal. On 15 February, during a night flight, Flying Officer John Gould undershot the aerodrome by half-a-mile, coming down in trees tearing off both of his fighter's main planes and the tail. The aircraft, A58-60 (QY-K), was a write off, but Gould only suffered a knee abrasion and shock. He was found fifteen minutes later wandering around with a deployed parachute in hand, although he didn't jump out of the aircraft. Realising his predicament too late, he had retracted the undercarriage but crashed before he could put on full power. Two days later, Flying Officer Ern Matthews "pranged" A58-52 (QY-S) while landing on the airstrip leaving him slightly injured. The aircraft went in for repair. He had swung

The partially camouflaged RAAF radar station at Cape Don.

off the runway into rough ground and the aircraft subsequently turned over on its back. He repeated the trick on 25 February sending A58-106 (QY-P) to the repair unit and as a result, he was posted out of the squadron two weeks later.

Flying Officer Bill Ford who was killed while flying A58-69 to Wyndham on 27 February 1943.

Next, on 21 February, Sergeant Paul Tully suffered an engine failure at 28,000 feet and glided down through cloud. On landing he turned the aircraft too far left, and it crashed at right angles to the strip. While Tully suffered abrasions, A58-22 (QY-K) was a write off. Then on Saturday 27 February, Flying Officer Bill Ford departed Strauss for Wyndham in A58-69 as part of No. 452 Squadron's defence duties that were required to cover the airspace over the west as well as the north. The alarm was only raised five days later when another flight from the squadron to Wyndham returned with the news that he was missing. After an extensive search, the burnt-out wreckage of A58-69 was discovered standing vertically in the ground at Litchfield Park, Northern Territory, and unfortunately Ford was still in it. His body was recovered and like so many, was buried in the Adelaide River War Cemetery, about 70 miles south of Darwin.

Flight Sergeant Eric Moore who was killed on 7 March 1943 when his aircraft crashed after an engine failure.

As if the squadron had not had enough grief, a week later Flight Sergeant Eric Moore died when his spitfire A58-98 developed a glycol coolant leak at 28,000 feet during an air test. The engine seized and as the weather was bad with total overcast, he came down to 1,000 feet in cloud with the intention of baling out, but he crashed about five miles from Strauss. He had left it too late.

It was a disappointing start and indicative of the relative inexperience of many of the RAAF's newer Spitfire pilots. The "old boys" from Europe were few and the novices had limited flying experience. Within the first three months, No. 452 Squadron had lost three pilots killed and eight aircraft either written off or in for major repairs.

By the time the Spitfires had arrived "up north", the Japanese had already sent fifty-one raids over Australia. The fifty-second on 2 March 1943 would be the first to be engaged by Spitfires. On this occasion, Nos. 54 and 457 Squadrons with Caldwell in the lead would be scrambled to intercept the raiders. The results were inconclusive with only a single Zero lost by the Japanese. No. 452 Squadron was not involved.

On 15 March 1943 the Japanese launched Darwin Raid 53. Comprising a flight of 22 Betty bombers supported by over twenty Zeros, the raid was met by No. 452 Squadron at 1100, but only six of the squadron's pilots were in a position to engage at 22,000 feet. Flying Officer Tim Goldsmith claimed a Betty bomber and a Zero and damaged another, while Flying Officer Ed Hall claimed he too had damaged a Zero. But the tables had turned. In what would be remembered as the squadron's first Australian action, they lost their commanding officer, Squadron Leader Ray Thorold-Smith, who was shot down into the sea. His body was never recovered but he is remembered on the Northern Territory Memorial at Adelaide River. Almost lost in the same engagement was Flying Officer Clive Lloyd who baled out of BS293 (QY-E) and landed safely west of Picinic Cove. It transpired the Australian claims were somewhat ambitious. Japanese records show only one (some records say two) Zeros were lost that day.

Squadron Leader Ray Thorold-Smith

Squadron Leader Ray Thorold-Smith who was shot down on 15 March 1943.

Ray Thorold-Smith had joined the RAAF in May 1940 and after initial training in Australia, was sent to Canada under the Empire Air Training Scheme where during pilot training he was assessed for fighters. He was commissioned on 17 May 1941 and sent to the UK, where he completed his Spitfire training before joining No. 452 Squadron as a pilot officer, the first Australian to do so. It would be during his time at Kenley that he soon showed his mettle when on 9 August 1941, he shared a Bf 109 with Finucane and added two more a fortnight later. September added to his score and in October he was appointed A Flight Commander.

On 22 November 1941, the King was graciously pleased to award Thorold-Smith the Distinguished Flying Cross for:

> ... showing aggressive spirit in destroying five enemy aircraft, sharing the destruction of another, damaging two and throughout, displaying skilful and capable leadership of his flight.

By the end of November, Thorold-Smith had "officially" recorded six destroyed, one shared destroyed and one damaged. Promoted to squadron leader on 18 March 1942, he assumed command of No. 452 Squadron after "Bluey" Truscott was posted out. He continued in this role, bringing the squadron home only to work up a new batch of pilots for the defence of Darwin in 1943. His loss on 15 March was keenly felt across not just No. 452 Squadron, but by the entire wing.

Thorold-Smith's replacement was Squadron Leader Ron MacDonald. MacDonald was already a veteran of the Pacific War. He had joined the permanent air force in 1936 and after pilot training in 1940 did tours with No. 67 Squadron, RAF, in Singapore, and No. 25 Squadron, RAAF. He commanded No. 12 Squadron, RAAF, before being appointed commanding officer of No. 452 Squadron. MacDonald inherited a squadron at full strength – twenty officers and ten NCO pilots and seven ground staff officers and 309 airmen. Aircraft strength was 23, which was probably a good thing given the number of accidents suffered by squadron pilots, who so far had managed to do more damage than the enemy.

Sergeant Bill "Lucky" Coombes who survived several incidents and later rose to the rank of group captain after the war.

In March, as well as Moore and Thorold-Smith killed, Sergeant Bill Coombes was slightly injured on 22 March when he swung on landing and ripped the undercarriage off QY-B (BS163). The Spitfire was written off, and Coombes got the nickname of "Lucky". After this crash he ditched into the sea in 1945 but was uninjured. He would go on to serve in the RAAF postwar, rising to the rank of group captain.

Spitfire Vc and A6M2/3 Zero

The Spitfire Vc was not well suited to operations in Australia. While a magnificent design which together with the Hawker Hurricane saved England during the Battle of Britain in 1940, flying in the tropics was a different matter. It was designed to operate in the much cooler and less humid European climate. Consequently, there were several problems when flying in Australia. First, the use of the Vokes tropical engine air filter on the Mk V Spitfires severely restricted the airflow into the engine which under humid tropical conditions led to underperformance and cooling problems. Engine failure, particularly at height became too common. Regarding the engines, Leading Aircraftman Arthur Brealey wrote in his diary:

> I detested the idea of working on the Merlin engines, but I could not help liking the way the boys used to fly them.

Second, the airfields in England were either long flat grass strips or made of concrete making take-off and landing easy. In the Northern Territory, they were graded dirt which caused engine problems with dust and stone ingestion, and they became muddy and rutted after heavy rains. The Australian bush each side of the landing strip made sure an off-course Spitfire would not escape damage. Third, the Spitfire's outwards folding undercarriage set close to the fuselage meant the slightest yaw might result in a ground loop with subsequent collapse or detachment of the undercarriage and the bending of the propeller – a problem for both theatres. The final problem was aircraft range. This was sufficient for short flights over the English Channel and across to France but was a severe limitation when flying from far south of Darwin to over Bathurst Island or the Arafura or Timor Seas.

The Spitfire's main opponent was the A6M *Reisen* series of fighters, commonly called the Zero. This aircraft was everything to the Japanese that the Spitfire was to the British and Australians. When it first appeared in the Pacific theatre, it came as a huge shock to the Allies who had expected a much inferior fighter flown by poorly trained pilots. To their surprise, the Zero performed better than the Kittyhawks and other types then in theatre and the Japanese pilots were well trained and experienced having flown in combat over China for years prior.

In the early war years, the Japanese Mitsubishi company had perfected the Zero design after a series of lesser-performing predecessors and much feedback from Japanese pilots. The early A6M2 variant was the model used over Pearl Harbor. To improve performance, it was not fitted with defensive armour plate or self-sealing fuel tanks at the insistence of pilots. The pilots felt their role was attack and preferred a lighter, more agile aircraft unencumbered by the extra weight this equipment would provide. This it proved to be, easily outmanoeuvring the Spitfire in a dogfight and given its extraordinary range and endurance, it could also outlast the Spitfire during longer combat engagements.

From September 1942, a new variant of Zero, the A6M3 appeared over New Guinea and Guadalcanal. It had shorter wings and better performance with a more powerful supercharged 1,130hp engine that led the Allies to initially believe it was a new type. It was given the code name Hap, but this was changed to Hamp to supposedly avoid embarrassing USAAF General Henry "Hap" Arnold. Then later this was changed again to the Zeke 32 when it was realised it was just a modified Zero. This and the A6M2 were the versions encountered over Northern Australia. Their technical specifications are in table below, in comparison with the Spitfire Mk Vc.

A58-2 was one of the original Spitfire Mk Vcs delivered to Australia in October 1942. It was issued to No. 452 Squadron in May 1943 and was flown by Pilot Officer Colin R Duncan with the squadron code QY-F.

Spitfire MK Vc and A6M2/3 Zero Technical Data

	Spitfire Mk Vc	A6M2 (Model 21) Zero	A6M3 (Model 32) Zero
Power plant	1,415 hp Merlin 46 V-12 liquid-cooled piston engine	950 hp NK1C Sakae 12 14-cylinder air-cooled radial	1,130 hp NK1F Sakae 21 14-cylinder air-cooled radial
Dimensions			
Span	32 feet 2 inches	39 feet	36 feet 1 inch
Length	29 feet 11 inches	29 feet 9 inches	29 feet 9 inches
Height	11 feet 5.5 inches	10 feet	11 feet 6 inches
Weights			
Empty	5,050 lbs	3,704 lbs	3,984 lbs
Loaded	6,650 lbs	5,313 lbs	5,609 lbs
Max	-	6,164 lbs	-
Performance			
Max Speed	374 mph at 6,000 ft	331.5 mph at 14,930 ft	338 mph at 19,685 ft
Cruising Speed	272 mph	207 mph	230 mph
Rate of Climb	5.6 min to 20,000 ft	7 min 27 sec to 19,685 ft	7 min 19 sec to 19, 685 ft
Service Ceiling	35,500 ft	32,810 ft	36,250 ft
Normal Range	470 miles	1,160 miles	-
Max Range	-	1,692 miles	1,284 nautical miles
Armament	4 x 0.303-inch MGs 2 x 20mm wing cannons	2 x 7.7mm MGs 2 x 20mm wing cannons	2 x 7.7mm MGs 2 x 20mm wing cannons

An A6M2 Zero of the 3rd Naval Air Group in September 1942. Shortly afterwards this unit was redesignated as the 202nd Naval Air Group and its Zeros were those faced by No. 1 Fighter Wing over Darwin in 1943.

An airman unplugs the starter battery from Spitfire QY-E prior to a sortie.

As well as protecting Darwin and the main road south, No. 452 Squadron began short deployments to Drysdale 370 miles to the southwest, and Milingimbi 310 miles to the east. Many assumed the arrival of No. 1 Fighter Wing in January 1943 meant that fighter cover could be provided across the whole of the Northern Territory. However, the vast distances involved made this impossible, so Nos. 452 and 457 Squadrons conducted regular detachments to Drysdale and Milingimbi more as a flag wave than with any expectation of hostilities.

From May 1940 Drysdale River Mission had a primitive 1,300-yard runway operating that had been graded from the bush. To this was later added a similar cross-runway about 1,110 yards in length. The airfield had been reconnoitred six times between March and November 1942 by either a lone Mitsubishi C5M Babs reconnaissance plane, or one Babs escorted by three Zeros – looking for Allied activity. No bombs were dropped, but a year later, a single Mitsubishi Dinah overflight on 26 November 1943 forewarned of an impending attack. That came the next day when 21 Kawasaki Lily twin-engined light bombers escorted by four Zeros bombed Drysdale and nearby Kalumburu mission stations. It would be their one and only raid on the remote location.

Milingimbi had been established as a coast watching station by the Department of Defence as early as 1927, run by aboriginal mission staff based there. By May 1941, the station was considered an Advanced Operational Base, but it was little more than a pair of dusty cross strips, one of 1,800 yards and another of around 1,600 yards. The Darwin station commander at the time noted the runways:

> ... were only clearings in which the surface was rough, in many cases sandy, and usually unserviceable after rain.

Fear of invasion forced the Directorate of Works to accelerate development, but the base remained austere. After the Japanese first raided the airfield and storage facilities in May 1943, No. 452 Squadron began rotations of three Spitfires if nothing to raise the morale of the small population living there. These rotations were shared with No. 457 Squadron. Milingimbi suffered four raids, the last of which was on 10 August 1943.

By accounts from ground personnel, Milingimbi not a popular posting. Helped by the local Yolngu aboriginals, fuel in 44-gallon drums dropped on the beach by boats had to be hauled up to the airstrip where the contents were hand-pumped into the Spitfire's eighty-five-gallon fuel tank – back breaking work in the heat and flies. Returning from one eight-day deployment, pilots Flying Officer Ed Hall, Flying Officer Clive Lloyd and Sergeant Paul Tully debriefed the intelligence officer on their return by stating we "encountered no hostile activity apart from mosquitoes and other insect pests". Milingimbi deployments continued into 1944.

While squadron scrambles became a regular event, most resulted from either false alarms triggered by the radar or an interception of a friendly aircraft – typically a straggling bomber returning from a raid or an aircraft in transit. As the war progressed and with the Japanese heavily occupied in New Guinea and the islands to the east, enemy raids over Darwin became less frequent. It would be almost two months before the next raid arrived overhead on 2 May 1943. As before, the raiders consisted of eighteen Betty bombers, with an escort of 27 Zeros.

Raid 54 began at 1015 and was met at 25,000 feet by Wing Commander Caldwell leading the wing up for attack. For this raid, the target was the RAAF station and harbour facilities. After a series of dogfights – some successful, others not so, No. 452 Squadron claimed a Betty and Zeke destroyed, a Zeke as probable and two Zekes damaged. Official wing reports claimed that five enemy fighters and a bomber were destroyed, four fighters were probably destroyed and nine fighters and a bomber were damaged.

In the thick of the action was Flying Officer Allen Mawer. His diary entry recalled:

> We rendezvoused over the strip at about 10,000 ft and formed up with the other two squadrons, making 33 aircraft. The assembled wing climbed to 31,000 ft over the top of Darwin and we saw the bombers, about 27 of them, heading for the town at 22,000 ft. They bombed the aerodrome but only damaged the end of the runway ... We trailed the formation on a parallel course, working our way up sun, and after what seemed like an eternity got into attacking position about 40 miles from the coast. 457 and 54 were sent against the bombers while we slammed into the fighters.
>
> After a long dive we reached them at 20,000 ft. I saw a big twin-engined bomber fall from the sky like a torch. It was a beautiful shot. The squadron engaged about three fighters, two of which singled me out for attention. Our little dogfight was terrific until I found that my cannons had jammed, and I was trying to deal with Nippon's finest with a few rifle calibre Brownings. Just when I thought I'd had it my No. 2 Dave Evans turned up.
>
> The extra Zeros he brought with him were less welcome, but we proceeded to attack the

Flying Officer Allen Mawer in his Spitfire QY-M.

> lot ... In one head-on pass I saw a lot of my rounds striking the Jap's engine and he passed under me with smoke pouring from it.

Mawer escaped despite most of the Spitfire's guns being jammed and he later claimed a probable victory.

However, the Japanese had more than evened the score. Thirteen Spitfires had been lost with one other damaged but repairable. Two pilots were killed: one was Flying Officer Alex McNab from No. 452 Squadron whose Spitfire A52-12 (QY-F) crashed into the sea about twenty miles off Perrin Island. His body was never recovered. Other casualties were Sergeant Ross Stagg who parachuted into the sea and Pilot Officer Ken Fox who also baled out. Sergeant Willie Nichterlein belly-landed. For the Japanese, five fighters and one bomber were lost.

For Ross Stagg, his adventure was only just beginning. Floating way out to sea, he somehow made it to shore, but it would be two weeks before he would be found. He attempted to walk the 60 miles back to Darwin while fighting to survive the heat and inhospitable countryside. He was eventually found near Mount Finniss suffering malnutrition and exhaustion and after a long spell in hospital, he returned to the squadron in Morotai in 1944.

It transpired that of the Spitfires lost by the wing, three ran out of fuel, three suffered engine failure and seven were lost due enemy action. Another who was shot down was Tim Goldsmith. Goldsmith was flying Spitfire A58-34 (QY-J) as callsign White Three and was ordered to scramble to intercept the raiders. He fired into a Betty bomber at close range but was in turn attacked by a Zero. His Spitfire was hit and flicked over, throwing him out of the cockpit and he parachuted into the water. He was rescued in his dinghy the next morning by a naval vessel and returned to the squadron. Goldsmith was a veteran of the Malta campaign becoming an ace there with twelve-and-a-quarter victories to his credit. He added this Betty to his tally. He would later add another three finishing the war with sixteen-and-a-quarter enemy aircraft confirmed shot down. It was not just Goldsmith that had success that day. Flying Officer Mawer claimed one and damaged one, Evans claimed a probable and Hall damaged another.

Flight Lieutenant Adrian "Tim" Goldsmith. A veteran of the Malta campaign, Goldsmith would become the RAAF's second highest scoring ace of WWII.

The loss of so many Spitfires in one engagement drew significant criticism, including from American General George Kenney who rather patronisingly said it would not have happened if American fighters had been used. MacArthur's Headquarters announced that the Spitfires had received "heavy losses", a politically naive announcement and an intelligence blunder in wartime. Despite wartime censorship, the press picked up on the "heavy losses" and that the Spitfires had been "worsted in the air fighting". Not surprisingly, the Japanese also picked up on the reference and used it for propaganda purposes.

The claims forced an inquiry as well as an official rebuttal in that:

> The heavy losses which the Spitfires suffered were not caused by enemy action or enemy superiority but by circumstances which were possibly accentuated by the very qualities which gave our airmen their victory.

However, the criticism continued which No. 452 Squadron commanding officer Ron MacDonald was forced to refute. A year later in a joint interview with Squadron Leader Ken James, the commander of No. 457 Squadron, MacDonald stated:

> That under no circumstances could it be said that the Japanese tactics were superior. The suggestion that the [enemy] drew us too far from our base is not true. Our boys were warned of a possible petrol shortage while in the air, but when some aircraft were shot down other pilots went looking for them and fixing their positions with the result that petrol supplies ran short and approximately seven of our aircraft landed on the beach. We lost only two pilots in the raid and most of the aircraft were salvaged.

Regardless of his upbeat assessment, No. 1 Fighter Wing did not perform well. Inexperience, poor Spitfire engine performance and lack of range all contributed to the circumstances. It was now clear that the Spitfire Mk Vs were no match for the Zeros in dogfights. The Zero could easily outturn the Spitfire, even one in an experienced pilot's hands. From May 1943 onwards, the wing was ordered to "bounce" the enemy, fire, dive and escape.

Spitfire accidents were fairly common. Flight Sergeant Arch Heaney was fortunately unhurt after QY-R swung on landing at night and hit an embankment at Sattler.

All was relatively quiet until 20 June when Raid 55 came in. This time, it was the Japanese Army's 61st *Sentai* based in Timor that made its first and only Darwin attack with eighteen Helen heavy bombers, escorted by 22 59th *Sentai* Oscar fighters. They were joined by nine 75th *Sentai* Lily light bombers. The raid was led by Major Takeo Fukuda the commanding officer of the 59th *Sentai* and had been preceded by a pair of Japanese reconnaissance aircraft three days prior. This time, intelligence forewarned the squadrons that the Japanese were concentrating at their Timor base, so by 0800 No. 1 Fighter Wing was at readiness. By 0950 the raid was underway with the RAAF aerodrome and stores the target. No. 452 Squadron made:

> ... an excellent interception destroying three probable Sally Mk 2 bombers one Zeke fighter and in addition damaging two enemy bombers and one fighter before they reached their target.

In fact, only two Helen bombers were shot down and an Oscar. Flight Lieutenant John Bisley as Blue One was one who claimed a "Betty" destroyed when in fact it was a Helen. Bisley had reached ace status in the skies over Malta where he was credited with six-and-one-half enemy destroyed and earned a DFC. The Helen he shot down would be his last claim for the war. His combat report read:

> I followed Red Section onto the bombers when they went in. Two Haps were 4,000 ft below and Blue Three asked permission to take in Blue Four and Five. My No. 2 and myself attacked the bomber force of 20 plus from the starboard beam. I attacked the first starboard bomber from a quarter beam at 300 yards closing to 50 yards from dead astern. The starboard engine of the bomber was burning when I broke off to evade enemy fighters whose tracer I observed passing close to my cockpit. As I broke violently

> away, I saw a bomber burning with both wings snapped off nearly up to the engine. This Betty spun into Adam Bay about five miles offshore.

Numerous claims were made in the heat of battle, so misidentifications were inevitable. For the effort, No. 452 Squadron lost two more pilots killed in the action. Pilot Officer Willie Nichterlein's time was up having spent less than three-and-a-half months at the unit. He was shot down into Adam Bay. Lost too was Pilot Officer Tom Ruskin-Rowe. As Blue Two he had only flown six flights with the squadron since joining, a testament to the levels of inexperience of the latest generation of RAAF pilots arriving in theatre. Ruskin-Rowe was shot down over Koolpinyah Station near Hope Inlet. Despite these losses, No. 1 Fighter Wing had gone some way to repair its damaged reputation after the disaster of 2 May.

Engine failures continued to bedevil the Spitfire Mk Vc. Flying Officer Reg Williams, one of the "originals" from Europe, was forced to bale out of A58-1 on 23 June and Flying Officer Gerald Cowell force landed on a beach five days later, the same day as Raid 56 came in. On this occasion, No. 452 Squadron did not engage the enemy. Williams was apparently quite relaxed upon landing despite his precarious situation. When daylight broke the next day, he was spotted safely on the ground, smoking a cigarette and sitting beside his parachute which was dangling in a tree.

Why had the Spitfires performed so badly over Darwin? Four reasons can be offered.

1. Engine Failures. In addition to the tropical conditions as previously mentioned, there was a serious problem with the Merlin engine's constant speed unit (CSU) or governor. This device automatically adjusted the propeller pitch to ensure the engine revolutions remained constant. CSU failures quickly led to an engine overspeed and eventual seizure and was to be the demise of many a pilot and aircraft. Yet another contributing factor was the Merlin liquid cooling system, so any leak in the glycol cooling fluid soon resulted in engine overheating and failure. This latter problem was easily detected as a streak of white smoke pouring from the engine cowl warning pilots of an overheat.

2. Gun stoppages were another problem that plagued the RAAF Spitfires. Pilots reported that often a gun would stop after only firing a few rounds or not work at all. This dangerous situation in combat often proved fatal. The Hispano cannons required warm air but even on a 30° Centigrade day on the ground in the tropics, the temperature at 20,000 feet was minus 25°C. This and the engine problems forced Air Commodore Frank Bladin, the Air Officer Commanding North-Western Area, to immediately request new Spitfires to replace the now worn-out aircraft under his command. No. 1 Fighter Wing had lost 44 Spitfires in six months of operations but claimed 63 enemy aircraft shot down. Time would show these claims to be an exaggeration.

3. Outdated attack formations. No. 1 Fighter Wing was using the three aircraft "Vic" formation and "Big Wing" tactics favoured by Caldwell. The "Vic" formation was a

three-aircraft formation in the shape of a "V" or rather a "Λ" with the lead aircraft at the apex. This was a standard RAF tactic going back to before the Battle of Britain. The lead aircraft, usually flown by a more experienced pilot, would lead the attack with his two wingmen in support. Once the lead pilot had finished his attack it would be the number two's turn and so on. The problem with this was the Zero could outmanoeuvre the Spitfire thus denying the lead a shot while the enemy could attack from the rear without being seen. The Germans and Japanese adopted the "finger four" formation – a four-aircraft astern flight with the lead either second in from the right or left depending on the lead's preference. The lead would be slightly ahead of the others similar to the four fingers of an extended right or left hand. This gave cover for the lead as well as covering the rear.

The "Big Wing" had been championed by the RAF's Douglas Bader and Air Marshal Trafford Leigh-Mallory but others in the RAF High Command thought it a poor tactic. The idea was to form up the squadrons so that they could attack the enemy *en masse*, but it took too long to form once a scramble had been called, and much fuel was wasted in doing so. In addition, the enemy could see the "Big Wing" approach and alter their tactics accordingly.

4. Lastly, the Spitfire had too short a range and endurance. Even with a 30-gallon drop tank fitted to the aircraft's belly, the Spitfire could barely manage 470 miles while the Zero could fly 1,160 miles. So, after flying out to meet the enemy over the sea northwest of Darwin, the Spitfire could not loiter or engage in combat for long before fuel became short while the Zero could fly from bases in Timor, fight and return with ease.

A nice shot of a Spitfire Mk Vc A58-43 (QY-A) which served with No. 452 Squadron from August 1943. During the 1943 Darwin campaign No. 1 Fighter Wing's Spitfires struggled in the face of engine and gun difficulties, poor endurance and unsuitable tactics.

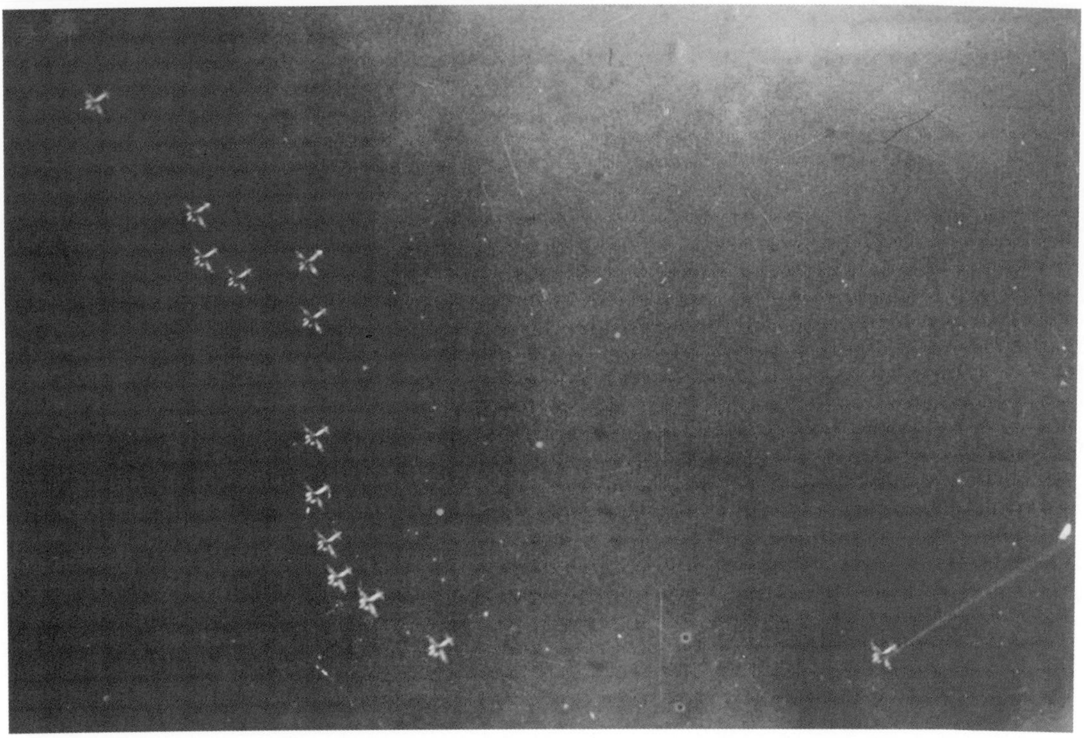

A formation of Japanese Betty bombers over Fenton, which was raided twice: on 30 June and 6 July 1943. One of the bombers has been hit and is lagging behind the formation trailing smoke.

CHAPTER 6

JULY TO DECEMBER 1943 · THE DAYS ARE NOT SO EASY

The three weeks from mid-June to early July 1943 would prove to be busy. Following Raids 55 and 56 (on 20 and 23 June as described in the previous chapter), Raid 57 arrived overhead on 30 June and Raid 58 on 6 July. Both raids targeted Fenton airfield. In May and June, the 380th Bombardment Group, USAAF, had arrived there with four squadrons of B-24 Liberator heavy bombers. Hence the Japanese interest in putting the airfield out of action.

Thus, when Raid 57 was detected, there followed the usual scramble and interceptions. On this occasion, the IJN launched 22 Betty bombers and 27 Zero fighters against Fenton. Misunderstandings meant that No. 452 Squadron's counterattack did not go well. An impressive total of 38 Spitfires were launched by No. 1 Fighter Wing, but No. 452 Squadron attacked the bombers and not the fighters as ordered. This allowed the Zeros free rein with the inevitable result, and despite the tactic Fenton was badly damaged by accurate bombing.

During the raid, No. 452 Squadron lost three aircraft and a pilot due to engine seizure. The Japanese lost none. The first of these was Flying Officer Bill Lamerton in Spitfire A58-19 (QY-T). Lamerton, another of the "originals" from Europe had a sad end. Following the scramble against the raiders, his engine suffered cooling trouble, so he returned to the airstrip. He overshot the runway and crashed. The Spitfire immediately caught fire, and he was trapped. Although rushed to hospital, he succumbed to his burns. Sergeant Keith Cross in A58-52 was luckier and survived his wheels up crash landing with only cuts and abrasions.

The third aircraft lost was flown by Flight Sergeant Col Duncan, who survived his parachute landing and later recorded his seven-day survival adventure. With his Spitfire engine on fire and after fighting to open his cockpit canopy to get out of his burning aircraft, he was finally able to extricate himself as the aircraft passed 24,000 feet in a left-hand downward spiral dive. He recalled being launched into the airflow at 300mph and hearing an explosive roar, which he thought was either the wind or the aircraft blowing up:

> I tumbled slowly over and over, revelling in that delightful icy air, away from the fury of the fire at last. I would have to do a delayed drop, so as to fall as far as possible below the combat area, for safety from being shot at. I knew that it took one-and-a-half minutes to free fall to the ground from 19,000 feet and I needed to get to a lower altitude for breathable air. I had both hands on the ripcord handle and tried to judge a one-minute delay, but it must have been a lot less, for I opened the parachute at about 15,000 feet, where it was spotted by another 452 pilot - Ross Richardson.
>
> ... eventually I crashed heavily into a gum tree due to a 20mph wind and finished up with my feet hanging only a couple of feet from the ground. I released my chute harness and dropped to the ground, I then waved to my guardian, who in response waggled his wings and flew off.

Flight Sergeant Col Duncan beside Spitfire Mk Vc A58-2 (QY-A), from which he baled out of on 30 June 1943. This same aircraft is featured in the profile on page 76, and can also be seen with squadron code "T" on page 89.

Duncan landed miles from nowhere. He knew his Mae West contained an emergency survival kit which he unzipped. He continued his story:

> There was a steel mirror about 4 x 5 inches for flashing. There was a small supply of a sort of chocolate that didn't melt in normal high temperatures and wasn't very tasty, milk tablets for sustenance, a supply of aspirin, mercurochrome, gauze bandages and a small tube of morphia, with an injecting needle and a plastic cover. There was also some mosquito netting and many safety pins and a booklet on how to survive if forced down. There were also some wax "vesta" matches in a tin and a compass. My right arm throbbed constantly so I broke open the loose hanging skin from my wrist to above my elbow, releasing the liquid and applied the mercurochrome to it so as to prevent infection for there were many flies and mosquitoes about. I then bound it up with the gauze bandage and safely pins and then rigged up a sling to support it to keep my hand high to reduce the throbbing. I gave the same treatment to my knees as they were also burnt.

For the next seven days, Duncan would await a ground party rescue but at least he had food and provisions dropped by various aircraft that knew his position. He would spend two-weeks in hospital before returning to the squadron.

Although suffering three losses, No. 452 Squadron recorded a Zero shot down by Flight Lieutenant Teddy Hall, one of the flight commanders. Hall was flying as White One and reported after unsuccessfully attacking the bombers his very close shave:

> A Zeke attacked me from nearly head on – firing white tracer at close range. I opened fire and closed to nearly 50 yards with a two second burst. I observed strikes and white and black smoke poured from his engine ... The Zeke turned on its back and went down out of control.

Despite his bravado, Hall would not make ace status claiming two-an-a-half aircraft destroyed, plus a probable and two damaged.

The Repair and Salvage Units and Aircraft Repair Depots

Repair and Salvage Units (RSUs) were large maintenance units comprised of up to ten to twelve officers and over 500 senior NCOs and airmen of all trades whose job it was to bring damaged and unserviceable aircraft back into frontline service. Sixteen such units were formed during the war.

Trades included specialists in airframes, engines, armament, instruments and electrical systems, in fact everything needed to completely rebuild an aircraft from parts. An air force policy of March 1942 delineated the responsibilities between the squadrons and the RSUs. Squadrons would be responsible for daily pre- and post-flight inspections and minor repairs, but anything beyond that required the work of the RSUs. As their name suggests, a major part of the RSU's work was salvage. This often meant taking trucks and cranes deep into the remote bush to recover a crashed aircraft, and in some cases a body.

No. 7 RSU and later No. 9 RSU became responsible for Spitfire maintenance "up north". No. 7 RSU formed at Richmond, NSW, and moved to Pell airstrip in January 1943. In February 1944, it moved from Pell to the Darwin civil aerodrome where there were hangars, more space and more civilised living conditions. At war's end, the unit disbanded in August 1945. As well as No. 7 RSU, from May 1944, No. 9 RSU also

Mk V Spitfires undergoing maintenance at a remote airfield before being returned to a squadron.

became responsible for Spitfire maintenance when it was established at Sattler. With the move of the Spitfire wing to Morotai in 1945, No. 9 RSU moved with it. When the war ended, No. 9 RSU received war expired Spitfires to prepare them for disposal prior to the unit disbanding in September 1945.

If the work was beyond the capability of the RSU, then the aircraft was sent to an Aircraft Repair Depot. These were called third line maintenance units as the RSUs were classed as second line and the squadrons first line. Four such depots were formed between 1942 and 1945 with No. 14 ARD located at Gorrie in the Northern Territory from February 1943 to manage Spitfire work. At this time, No. 14 ARD had thirteen officers and 135 airmen.

At least nine of No. 452 Squadron's Spitfires ended up at No. 14 ARD, particularly those that had belly landed, had their undercarriage collapse or had ground prangs. As well as Spitfires, the depot also repaired Mitchells, Hudsons and Beaufighters. From February 1944, No. 14 ARD moved forward to the Darwin civil aerodrome only to end its days there in August 1945 when the war ended.

At the end of the war, the RAAF had little need for its Spitfires, so a place was needed for storage pending disposal. No. 6 Aircraft Depot at Oakey near Toowoomba was chosen for fighter aircraft which included Boomerangs, Mustangs, Kittyhawks and Spitfires. A total of 544 aircraft awaited their fate, 232 of which were Spitfires. Sadly, for modern military aviation enthusiasts, that fate was scrapping.

By now, life had settled into a familiar routine. A Japanese raid would come in, the Spitfire wing would respond and after tallying their claims and their losses they would await the next raid. Unfortunately, losses would exceed claims. Raid 58 on 6 July would well illustrate the wing's problems. While claims continued, three in this case, losses outnumbered them. Not that it was a game of attrition, more of a morale sapping cycle. The raid consisted of 22 Bettys and 27 Zeros and again, Fenton airstrip was the target. The Japanese lost two Bettys although No. 452 Squadron claimed a Betty and a Zeke, with another damaged. For the squadron, it was the same old story – two engine failures and one Spitfire shot down, with Flying Officer Clive Lloyd the lucky one parachuting safely down. Lloyd was to be joined on the ground by Sergeant Ross Richardson whose engine had overrevved, and he was also forced to bale out.

The third casualty was Flight Lieutenant Paul Makin, an RAF officer on exchange who had been with No. 452 Squadron since the days at Kenley. He had been appointed flight commander and after Thorold-Smith's loss, was among the last remaining "originals". He was later found safe and briefly returned to the squadron until late July.

On 10 July, No. 452 Squadron had an interesting team visit them from "down south". Led by Professor Frank Cotton of Sydney University, and under the direction of Flight

Lieutenant Ken Roberston, the squadron was to trial the Cotton Aerodynamic-Anti-G suit or CAAG. Several countries were experimenting with G-suits including the Canadians, British and Americans to improve human response under "G" force conditions such as are experienced by fighter pilots. The human body can "blackout" or go unconscious after sustained high "G" forces over +6G or six times the force of gravity. The suits used an inflatable bladder to compress the legs and body to stop blood pooling in the lower extremities, thus maintaining a blood flow to the brain.

Flight Lieutenant Ken Robertson who tested the first G-suits for the RAAF.

Cotton's invention was seen by the RAAF hierarchy as a possible counter to Japanese superiority in aircraft performance, particularly in tight turns. They surmised that the suits might permit pilots to out-manoeuvre the Zero in dogfights. The Zeros exceptional manoeuvrability meant that the response was not to engage in a dogfight but to rather use the Spitfire's speed to attack quickly and then dive away. After testing the suits in a centrifuge at the University of Sydney, Robertson first flew it in a Hurricane Mk 1, the only one in Australia. He eventually buckled the aircraft's wings at +10G. Further tests on Spitfires and Kittyhawks proved promising, but its use in a Boomerang aircraft was discontinued. The trials on an operational squadron were set for September and October 1943 and No. 452 Squadron was chosen. Robertson would eventually be posted to the squadron in November. For his work on G-suits, he was awarded an Air Force Cross in September 1943. The citation read:

> Flight Lieutenant Robertson has carried out the flight trials of a device to improve the fighting qualities of fighter aircraft by extending the factor of human endurance and has been responsible for its gradual refinement until it has reached a form suitable for operational use.
>
> Throughout these tests Flight Lieutenant Robertson has submitted himself, without hesitation, to strains which are the limit of human endurance, thereby displaying outstanding courage and devotion to duty.

A quiet period between July and August allowed for continued experiments on the suits which proved successful and after refinements, six sets were installed in six No. 452 Squadron aircraft. Caldwell was one to laud the suits but noted the aircraft could be easily overstressed. The pilots called them "zoot suits" and by 1 October 1943, the entire squadron was kitted out.

Spitfire A58-84 with the unusual squadron code "QY-?". Other odd symbols used by No. 452 Squadron were a delta from the Greek alphabet and a plus sign.

The pilots were initially enthusiastic but doubts soon arose. The suits were cumbersome in the tight cockpit making egress difficult and the suit equipment added weight. Many worried the suits would drag them down if they had to swim after coming down in the water. Nevertheless, pilots began to wear them, although it appears the only time they were used in combat was when Flying Officer Ken Fox attacked an enemy escorted reconnaissance formation of 36 fighters on 7 September 1943, but without result.

Using G-suits or no suits, the war went on. The quiet period was broken on 10 August when Flying Officer "Red" Young and Pilot Officer Bill Coombes caught what they called two "Pete" floatplanes off Millingimbi and destroyed both. In fact, they were a Jake and a Rufe, both floatplanes.

Meanwhile, with pressure now on to support MacArthur's New Guinea campaign, North-Western Area's role changed to one of supporting the American push by pinning down the Japanese to the west of New Guinea so they could not interfere. Consequently, US Fifth Air Force commander General George Kenney ordered the RAAF to increase a show of force in the area. Major targets would include Koepang (in Timor) and the Celebes. The USAAF build up had also begun.

As well as the USAAF, the RAAF made its own changes in the Darwin area. In mid-November, No. 1 Fighter Wing was disbanded with all fighter operations to come directly under North-Western Area for administration. Operational control would be exercised by No. 5 Fighter Sector. Recent experience had found that there was confusion as to who controlled whom. Months later, air force headquarters reintroduced the term No. 1 (Fighter) Wing, but the command hierarchy did not change.

While the RAAF reorganised, the Japanese also made changes to their dispositions once they realised the Allied counterattack would come in New Guinea. IJA air regiments of their

7th Air Division moved from Ambon to Wewak on 10 August while the IJN's 23rd Air Flotilla headquarters remained at Kendari in the Celebes. Their bases were at Penfoei (Timor), Ambon and on the Kai Islands, south of Dutch New Guinea. This naval force would be left to attack Australia.

A Japanese reconnaissance on 7 September 1943 by two *Dinahs* was escorted by 36 Zeros and Hamps. They were met by an equal number of Spitfires. The No. 452 Squadron unit historian recorded that on this occasion they were "jumped" badly. Again, the claims exceeded the losses and again the squadron's identification of the enemy was poor. Flying Officer Adrian Goldsmith claimed a "Tony", while Flying Officer Kenneth Bassett, Flight Sergeant Jeffrey King and Flying Officer Gerald Cowell each claimed a damaged "Oscar". The recognition of Pilot Officer William Coombes was more accurate as he claimed a Zeke as a probable. In fact, only one Zero was lost. Meanwhile, Squadron Leader Ron MacDonald and Pilot Officer Tully were both shot down and baled out. Tully escaped relatively unharmed, but MacDonald suffered burns from a shell that exploded in the cockpit. He later claimed the G-suit prevented his more rapid escape from the cockpit and thereafter showed a great deal of antagonism toward them. MacDonald's view obviously carried some weight as on 9 November, RAAF Northern Command advised RAAF Headquarters that wearing of the suits would be discontinued.

Little did the No. 452 Squadron members realise that this would be the last combat flying they would do in 1943. Japanese raids had by October all but ceased with the last, Raid 64 on 12 November, not engaged by the squadron. It would also be the last bombing of the Northern Territory. It was now readily apparent that any Japanese threat to land troops in Australia had all but disappeared.

Although the enemy had gone, accidents continued and on 26 September inattention claimed two more pilots. Pilot Officer John Adam and Flying Officer Allen Mawer collided in mid-air during a practice dogfight. The pair were two of a four-ship formation tasked to carry out formation flying, army cooperation and camera gunnery. At about 0800 and flying around 2,000 feet over the Manton Dam, Adam made a mock attack out of the sun towards Mawer. He closed well in and then pulled to starboard. Unaware of Adam's presence, Mawer commenced a steep climbing turn to starboard. His port wing struck Adam's port wing, and both wings broke off about six feet from the tips. Both aircraft commenced to roll out of control, before crashing into the bush and bursting into flames. Both pilots were killed instantly. Adam and Mawer were the last casualties sustained during No. 452 Squadron operations in 1943, but not the last to be killed. Squadron members Flight Lieutenant Al Watkin and Leading Aircraftman Stanley Kirk Sims died when a USAAF C-47A Skytrain (Dakota) in which they were passengers heading on leave crashed at Rewan Cattle Station in Queensland during a violent thunderstorm. All nineteen aboard were killed.

With the reduction in combat, many of the "old hands" who had completed their tropical tours began to be replaced. Flight Lieutenant Ken Bassett was one of them and typifies their war experience.

A Spitfire Mk Vc taxies past the viewing gallery at Strauss.

Flight Lieutenant Ken Bassett

Flight Lieutenant Ken Bassett (left) relaxing over a cup of tea with Flight Lieutenants Paul Makin (RAF) and John Bisley.

Born in Roma, Queensland, in July 1916, Ken Bassett enlisted in the RAAF in October 1940. He graduated as a sergeant pilot in May 1941 and was commissioned a year later. Showing aptitude to fly fighters, he was posted to England to undertake a Spitfire conversion course at one of the numerous operational training units, in this case No. 61, which was at Heston. The airfield is now Heathrow Airport.

With 88 Spitfire hours in his logbook, Bassett was posted to No. 452 Squadron in October 1941 which had just arrived at RAF Station Redhill in Surrey. A line pilot, he would share the duties and the scramble that the RAF fighter squadrons were experiencing. He did not shoot down any enemy nor crash but survived flying another 195 hours. This was remarkable.

When No. 452 Squadron was ordered to move to Australia in June 1942, Bassett went with the unit and upon arrival, found he would remain with it. After awaiting aircraft, the squadron arrived in Darwin in January 1943, but he had to wait until February before he got airborne again. Offensive patrols, convoy protection and responding to false alarms became his lot until Raid 58 came in. That day, 7 September 1943, would be Bassett's first and last combat engagement.

On 6 November 1943, Bassett's operational tour had ended, and he was posted south. By now he had completed 54 operational sorties and amassed 371 Spitfire flying hours. Given his experience, he became a flying instructor and joined No. 7 Elementary Flying Training School at Western Junction in Tasmania. Many instructors were later to say that flying with ab initio students was more dangerous than facing the enemy. Bassett survived, and by the end of 1944, he decided he had done his bit and requested his discharge. He wanted to resume his civil employment back in his home state of Queensland and start a family. On 14 February 1945, his appointment was terminated. Bassett died at 89 years old in May 2006.

For his war service in Europe and the Pacific, Bassett was awarded five medals: the 1939-45 Star, the Aircrew Europe Star, the Pacific Star, the War Medal 1939-45 and the Australia Service Medal 1939-45. He would also wear his Returned from Active Service Badge at parades and special occasions.

As the year came to a close, the Japanese were essentially a spent force in the Netherlands East Indies. Although they were still able to intercept Allied bombers they generally kept close to their home bases. Heavy losses of their forces at Rabaul and in the Solomons forced the 23rd Air Flotilla to release squadrons to face the US Navy in Truk and the Marshall Islands. Meanwhile, more changes were happening in the RAAF command as well. Wing Commander Caldwell had been replaced by a fellow ace from the Middle East, Wing Commander Peter Jeffrey, on 26 September. By this time No. 452 Squadron had 30 pilots and 362 ground personnel on strength with 24 Spitfire Mk Vcs on unit establishment. For all this, there was little to do apart from practicing scrambles and formation flying. Headquarters North-Western Area now turned their attention to the offensive.

Although No. 452 Squadron faced a lengthy quiet period after engaging the enemy on 7 September 1943, the flying and accidents continued. This photograph shows the result of a take-off accident by Flight Sergeant Arthur Richardson on 2 October 1943. The pilot was unhurt and the aircraft was returned to service.

A formation of No. 452 Squadron Mk VIII Spitfires, which the unit began receiving in January 1944.

CHAPTER 7

JANUARY TO DECEMBER 1944 · INTO THE FRAY AGAIN

The new year of 1944 would offer a nice surprise for No. 452 Squadron's pilots. Five of the long-awaited Spitfire Mk VIIIs arrived in January to replace the Mk Vcs but within days, the Mk VIIIs were grounded. The Mk VIIIs had much higher performance needed to gain the advantage over the Japanese Zeros they faced. The problems were with the engine oil system which were fixed by February. In time, all the Mk Vcs would be replaced by Mk VIIIs. Alongside the transition to new aircraft, leadership also changed within the squadron. Squadron Leader Ron Macdonald, who had served with distinction despite suffering burns when he was shot down, stepped down after nearly a year in command. His successor, Squadron Leader Lou Spence, officially took over on 3 February 1944.

Squadron Leader Lou Spence

Squadron Leader Lou Spence in his Spitfire A58-254 (QY-V) Rima II. Spence would serve as CO between February and November 1944.

Lou Spence was one of the RAAF's finest. He was already a veteran of WWII and would later serve in Korea. He had graduated from flying training in August 1940 and was posted to No. 25 Squadron in Perth flying the obsolete Wirraway before a posting to the Middle East where he joined No. 3 Squadron, RAAF. Flying a Kittyhawk on 26 January 1942, he landed to rescue fellow squadron pilot Sergeant Walt Mailey who had been shot down by flak. Spence was awarded the DFC, not so much for this action, but for displaying "fine leadership throughout [and] has destroyed two enemy aircraft". A bar to the DFC would be awarded posthumously for his leadership in Korea.

Following his tour in the Middle East, Spence returned to Australia to instructional duties at an operational training unit before being posted to No. 452 Squadron as commanding officer. He spent most of 1944 leading the squadron before handing over command in November. With the war over, his appointment was terminated in late 1945. But that was not the end of Spence's RAAF career. Short of pilots at the end of the war, he was reappointed to the Citizen Air Force in 1946 and was later offered a permanent commission. In February 1950, he was appointed the commanding officer of No. 77 Squadron then flying Mustangs in the Korean War. While leading a sortie in September, he was shot down while attacking ground targets at An'gang-ni, where his Mustang was last seen diving steeply into the ground.

Spitfire Mk VIII A58-254 Rima II as flown by Squadron Leader Lou Spence.

The early months of 1944 were uneventful. The Japanese raids had ceased and apart from detachments to Drysdale and Milingimbi, little serious flying was done. Boredom was relieved by making mock attacks on Liberators returning from their bombing missions. Morale was not helped by the heavy rains of the wet season which bred mosquitoes. A new arrival to take over B Flight, Flight Lieutenant Ron Cundy, recalled that "although we spent many hours every day on readiness (mostly playing cards) no action eventuated". He continued:

> The monotony of sitting around the flight hut in readiness waiting for the Japanese who never came was relieved from time to time with line-astern chases, battle climbs in squadron formation, shadow shooting and aerobatics.

No. 452 Squadron Spitfires make mock attacks on a returning RAAF Liberator bomber.

Cundy was another veteran of the Western Desert, having served with the RAF on Hurricanes and Kittyhawks. During that posting he became an ace claiming five victories, plus two shared, two damaged and one probable. He survived the war but would not get the chance to add to his score in the Pacific.

Flight Lieutenant Ron Cundy, another ace from the Western Desert who joined No. 452 Squadron in November 1943.

On 8 March, Spence got an urgent call to immediately proceed to wing headquarters. No. 452 Squadron was ordered to deploy to Pearce in Western Australia and plan on having enough spares to last ten days. Intelligence reports suggested a Japanese attack on the Fremantle naval base was imminent. Apparently, a US submarine had reported a Japanese naval task force sailing through the Sunda Strait between Java and Sumatra, and the assumption was it was headed for Perth.

Subsequently, between 9 and 24 March 1944, No. 452 Squadron deployed to Guildford airfield in Perth rather than Pearce under *Operation Potshot*. Hasty preparations were made and at 0830 on 9 March and despite a dire weather report, eight Spitfires departed Strauss for Derby accompanied by a Dakota full of ground crew. This trip would be long remembered, not for the threat of the Japanese but for the tropical weather.

They encountered heavy rain, low cloud and strong winds, so they elected to land at Wyndham. One Spitfire Mk Vc, JL630 (QY-D), flown by Sergeant Ron "Freddy" Fiveash missed the runway and landed in the mud where it nosed over. Fortunately, he was rescued by fellow pilots and taken to hospital. He returned to the squadron two weeks later. Five aircraft arrived at Derby the next morning with the other two a day later. Then it was on to Port Headland where they were met by a sandstorm and then heavy rain during the refuelling. The next stop was Carnarvon.

Still in poor weather, much of the next leg was flown out to sea to avoid the cliffs and rising ground. At Carnarvon, Sergeant Colin Beech crash landed with undercarriage trouble, but both pilot and aircraft were not badly damaged. After refuelling in a howling gale, the squadron set off for Perth where Flying Officer Col O'Loughlin force landed BR537 (QY-A) at Gin Gin with fuel problems. The unit historian was later to write "These aircraft were not intended for flights of this character". In fact, it was an epic journey of over 2,000 miles in appalling weather, and it was amazing that no one was lost.

To the ground crew's credit, five hours after landing in Perth, six of the seven Spitfires were ready for combat. Once at Perth, the squadron remained on alert until intelligence advised of a waning of the danger and the pilots stood down. After all the drama of the outbound flight, the return was uneventful.

Refuelling a Spitfire at Sattler in preparation for a sortie. Two interconnected fuselage tanks held 85 imperial gallons.

A group photo of No. 452 Squadron pilots in early 1944. Standing rear, left to right, are Toby Jones, Doug Hoile, unknown, Squadron Leader Ron McDonald, Fred Johnson, Ken Robertson, Lou Spence, Peter Bullock, Col O'Loughlin, Dave Murray, Keith Gamble and unknown. Squatting front, left to right, are Fred Fiveash, Ray Job, Bill Stevenson, Eric Mahar, Paddy Beeston and two unknown pilots.

Upon their return, No. 452 Squadron found little had changed. They stood at readiness, flew practice missions and waited. Possibly to keep up morale and relieve boredom, on 18 April, three Spitfires went on the offensive, the first time for the squadron since they had arrived in Australia. They were tasked with staffing an enemy wireless station at Tepa on Barbar Island in the Moluccas, a mission that was carried out with great success.

With a paucity of flying, maintaining skills became paramount. Scrambles to intercept returning friendlies or unidentified radar plots became regular events, but one flight ended in tragedy. On 24 April, Sergeant Colin Dunning in A58-232 was flying in a training exercise as Blue Two with Flying Officer Eric Mahar as Blue One. Somehow during the flight, Dunning's propeller clipped Mahar's and his engine failed. His aircraft descended and he went missing near Point Patterson. A wide search failed to locate the aircraft, and his location remained a mystery until his remains were finally located by local Aboriginals in 1947. Dunning received a service funeral and was laid to rest at the Adelaide River War Cemetery.

The airstrip at Sattler is the closest to Darwin and is at the 20-mile post south along the highway. Given the constant threat of a Japanese invasion, work to construct the strip was not commenced until late 1942. It is named after Flight Lieutenant Geoffrey Sattler, the pilot of a No. 13 Squadron Hudson who with his crew was killed in the Celebes on 12 January 1942. By 1944, airfield construction squadrons had laid a sealed runway of 1,800 yards length with gravel taxiways. A few buildings had been constructed to support flying operations.

No. 452 Squadron began to arrive at Sattler on 3 May 1944 and was followed by No. 457 Squadron on 12 June. The sister squadrons operated from opposite ends of the strip, with No. 452 Squadron to the north. Such an arrangement meant careful planning between squadrons to ensure runway collisions were avoided. By September, three squadrons of P-40 Kittyhawks had arrived making Sattler an extremely congested and busy airfield.

While the three squadrons of No. 1 Fighter Wing awaited the next Japanese raid, the RAAF formed No. 80 Wing on 31 March 1944 in preparation for an advance into the Netherlands East Indies. By May, the wing headquarters was also at Sattler. Although planned for mid-1944, the move into the islands would not occur until the end of the year. In time, Nos. 452, 457 and 79 Squadrons would transfer to No. 80 Wing leaving No. 54 Squadron, RAF, and two new RAF squadrons, Nos. 548 and 549, in Darwin for rear area defence.

Baby faced Pilot Officer Malcolm "Junior" Beaton aside his Spitfire A58-236 (QY-G) named Junior and adorned with artwork of a pilot wearing nappies. The aircraft went to No. 457 Squadron in July 1944.

Then finally some action occurred, the

first for 1944. No. 452 Squadron was fortunate to be on readiness during daylight hours on 12 June when a scramble was called at 0700. Twelve aircraft quickly got airborne to intercept a Japanese reconnaissance aircraft headed for Fenton. Of the squadron's twelve Spitfires, three engaged and shot the Dinah down into the sea, each claiming a one-third share. Flying Officer Gamble called the "Tally Ho" at 28,000 feet and made a successful starboard quarter attack which left a large hole in the Dinah's starboard wing and blew off the engine cowl. The Dinah then began a slow descent then Flying Officer Col O'Loughlin and Pilot Officer Malcolm "Junior" Beaton made their attacks. The Dinah glided into the sea and exploded. However, the victory came at a cost when O'Loughlin's engine overrevved, and he was forced to bale out into the sea. Fortunately, he was quickly picked up by a Catalina none the worse for wear.

Following the excitement of the Dinah shoot down, monotony again set in. Practice scrambles, army co-operation exercises, sporting competitions between units and bush hunting trips became the norm. Attesting to the boredom, in his monthly report of July 1944, the unit historian wrote:

> Morale is a very big factor up here these days with the war a long way off and no prospect of action from day to day. All we do is pray for the time to come when we will be told to get packing and see some war. There are a multitude of rumours as to the destination of the squadron and the cooks and hygiene squad as usual can be relied upon to be the first with the "REAL GOOD GUTS". Men with bright personalities are precious jewels these "Dog-Days" and keep others from brooding on the slow effluxion of time.

The only good news in July was the last of the Mk Vc Spitfires left No. 452 Squadron for disposal. The squadron would now own 23 Mk VIII Spitfires with 28 pilots to fly them. Reports stated that the new planes were "splendid". It transpired that the Mk VIII Spitfires

Spitfire Mk Vc A58-20 (QY-G) being repaired by No. 7 RSU in 1944 after it was ground looped to avoid a collision with another Spitfire. In July 1944 No. 452 Squadron disposed of its last Mk Vc Spitfires.

were well suited to tropical conditions after the earlier engine problems had been fixed. Equipped with a turbocharger that activated above 8,000 feet, the fighter had a much better rate of climb and high-altitude performance. These aircraft would also be fitted with G-suit equipment and Flight Lieutenant Ken Robertson returned to the squadron to supervise fitment and training. After unsatisfactory G-suit trials on the Mk Vcs a year prior, improvements had been made to pilot comfort and ability to egress the cockpit.

Spitfire Mk VIII and A6M3 Zero Technical Data

	Spitfire Mk VIII	Zero A6M3 (Model 32)
Power Plant	1,565 hp Merlin 60 / 70 12-cylinder liquid-cooled piston engine	1,130 hp Nakajima NK1F 21 14-cylinder air-cooled radial
Dimensions		
Span	36 feet 10 inches	36 feet 1 inch
Length	31 feet 3 inches	29 feet 9 inches
Height	12 feet 7.75 inches	11 feet 6 inches
Weights		
Empty	5,805 lbs	3,984 lbs
Loaded	7,807 lbs	5,609 lbs
Max	8,021 lbs	-
Performance		
Max Speed	404 mph	338 mph
Cruising Speed	324 mph	230 mph
Rate of Climb	4,800 ft/min	4,517 ft/min
Service Ceiling	41,500 ft	36,250 ft
Normal Range	660 miles	-
Max Range	1,180 miles	1,477 miles
Armament	4 x 0.303-inch MG 2 x 20mm cannon	2 x 7.7mm MG 2 x 20mm cannon

Flight Lieutenant Ron Cundy's Spitfire Mk VIII A58-435 (QY-T) named Pegasus. The aircraft was destroyed in a mid-air collision with a USAAF B-24 Liberator on 18 September 1944 when flown by Flying Officer Keith Kelly.

Egress was not an option for the next fatal accident suffered by the squadron. On 18 September, Flying Officer Keith Kelly flying Spitfire A58-435 (QY-T) had a head-on collision with a USAAF B-24 Liberator of the 380th Bombardment Group near Cape Van Dieman. Kelly was conducting fighter affiliation exercises when his port wing hit the Liberator's outside port engine. Watched by others in the formation, Kelly made no attempt to bale out, and his Spitfire was seen to crash. Nothing of it was later found. The Liberator was more fortunate, making a safe three-engine landing with no casualties.

One who doubly felt this accident was Ron Cundy. Although he had just been posted south, he later learned that also lost was his immaculate Spitfire *Pegasus* (QY-T), which his ground crew had polished to a mirror finish and which he later claimed was "the best aircraft I had ever flown". Pilot inexperience was to blame as Kelly had joined the squadron just six days prior and only had air-sea rescue experience albeit on Spitfires.

For the rest of 1944, no operational flying or scrambles were flown, but No. 452 Squadron was still beset by tragedy. While searching for Leading Aircraftman Maurie Wise, an airman who had gone missing and was later found safe, Flying Officer "Buster" Etherington as pilot of Wirraway A20-633 with Flying Officer Tom Smith as his observer crashed into scrub fifteen miles east of Sattler on 20 November while circling the ground rescue party. The aircraft was destroyed and both were killed.

No doubt this news greeted the new commanding officer, Squadron Leader Arthur Birch, with some consternation as he took over two days later. Birch was an experienced fighter pilot having flown with No. 54 Squadron, RAF, in the UK followed by Nos. 76 and 79 Squadrons in the Pacific theatre. Postwar, Birch would be awarded the Air Force Cross for his airmanship in South-East Asia on Dakota operations years later, but that is another story.

The monotony was broken and boredom turned to excitement in early December when news came through from headquarters that No. 80 Wing was to move to Morotai, an airfield in the Halmahera group of islands which had been invaded by Allied forces three months prior. Morotai in time would become one of the RAAF's most important bases. No. 452 learnt that they would lead the No. 80 Wing push followed five days later by No. 457 Squadron. Finally, No. 79 Squadron would make the move to bring the wing to full strength. To increase range, the Spitfires had to be fitted with long-range fuel tanks to make the trip. One tank sat under the belly containing 87 gallons and two under each wing held 25 gallons apiece.

The first eight C-47 Dakota transport aircraft arrived on 8 December to move the squadrons and preparations were made for a departure three days later. Packing, cleaning and site clearance was now the workload.

The move was not direct to their destination. The Dakota flights transited via Merauke in Dutch New Guinea, Noemfoor in Biak and then to the Pitoe airstrip on Morotai. Upon arrival at their new home on 12 December, No. 452 Squadron found the advance party and No. 60 Operational Base Unit had prepared the camp's operational areas and accommodation. Operational base units replaced station headquarters in 1942 and were established after a review of the debacle of the bombing of Darwin in February that year. The varied roles of these units included airfield management and repair, fuel storage and aircraft refuelling, fire response, local area communications, domestic and logistic services and transient aircraft maintenance. Airfield guards, medical staff and bomb disposal tasks were added to the list once the wing moved forward.

Prior to departure, Group Captain Caldwell called all the pilots together and gave them strict instructions. No pilot was to land short. On touchdown, all pilots were to open throttles to exit at the end of the strip. No pilot was to turn off onto a taxiway early. The advice would prove prescient. As a stream of Spitfires would land one after the other and be low on fuel, a blockage mid-strip would not only slow things down but also be dangerous.

No. 452 Squadron's 24 Spitfires then departed on 17 December via Gove but their arrival at Morotai ended in disaster. Flight Lieutenant John Sturm was one of four pilots "borrowed" from No. 457 Squadron for the movement. However, as he landed, he disobeyed orders and turned off early thus slowing his aircraft considerably. While Sturm was still on the runway, Flying Officer Bert Lumley who followed and as directed opened his throttle, tipped Sturm's rudder causing the aircraft to swing at right angles, but remaining on the runway. Lumley's aircraft lost its undercarriage and slid off the runway. He ended up in hospital with burns and lacerations and would not remain with the squadron. Next to land was Pilot Officer

A No. 452 Squadron photo to commemorate the November 1944 change of command. Seated front, left to right are the incoming commanding officer Squadron Leader Arthur Birch, Group Captain Clive Caldwell (the officer commanding No. 80 Wing) and the outgoing commanding officer Squadron Leader Lou Spence.

Bill Chrystal who landed flapless due to a failure. That meant his landing speed was higher than normal to prevent a stall. He collided with Sturm's now stranded Spitfire killing him instantly as the aircraft burned. Chrystal was also seriously injured and like Lumley, never returned to the squadron. All three had only recently been attached from No. 457 Squadron. They were members of No. 452 Squadron for only eight days.

Pilot Officer Bill Chrystal, a No. 457 Squadron pilot attached for the flight to Morotai who was involved in a landing accident in A58-325 on arrival. Chrystal was seriously injured and never returned to No. 452 Squadron.

Upon arrival at Pitoe, the squadron found the strip was of coral and hardstands had been bulldozed from the surrounding jungle. Also awaiting their arrival was a squadron of B-24 American Liberators and the occasional Japanese Betty bomber that came over at night which dropped a few bombs and left.

Although the RAAF provided Dakotas to airlift No. 452 Squadron to its new home, most of the ground crew and stores arrived by ship. The *New Mexico*, a passenger ship converted to a troop carrier, took until 1 February 1945 to deliver its passengers and cargo which also included men and materiel of Nos. 457 and 79 Squadrons.

The wing's role was now to clean up pockets of Japanese resistance throughout the islands and to prepare for the retaking of Borneo. All felt this was a complete waste of personnel and machines. The Japanese by this time were cut off from their supply routes and would eventually starve. Despite requests from Caldwell and the Americans to air force headquarters in Melbourne for the squadrons to move to the Philippines, all such requests were denied.

The Japanese were certainly aware of No. 452 Squadron's arrival. On 24 December, as if to welcome them with an early present, on Christmas Eve they sent two Nicks over to bomb the field. The Nick was the Allied code-name for the Kawasaki Ki-45 *Toryu* (Dragon Killer) IJA two-seat fighter. The anti-aircraft artillery and searchlights opened up on the first Nick without success. When the second Nick arrived overhead, Flying Officer Jack Pretty ignored the ground fire and took off and shot it down. Despite all this, Squadron Leader Doug Beattie was also airborne at the time. He was too far away to engage the enemy aircraft but was shot at by the trigger-happy gunners and returned with holes in his aircraft. It is not recorded what he later said to their commanding officer.

Despite the jubilation, earlier that day No. 452 Squadron lost another pilot, Flying Officer Ian Markwell. Markwell's Spitfire crashed into the sea after an engine failure following take-off.

He was flying a sweep led by Group Captain Caldwell. Despite a search, only his parachute was found. Markwell's loss was perhaps fated. He almost came to grief six days prior when he crashed at Merauke and ripped off his port oleo when he collided with other Spitfires.

1944 ended with an occasional intruder, but there was little further action, only frustration.

No. 452 Squadron Spitfires at Morotai, where they arrived in December 1944, with USAAF B-29s behind them. The large American bombers only used Morotai as a staging base.

A58-354 showing off the clean lines of the Spitfire Mk VIII and the tropical air filter fitted to the nose.

CHAPTER 8

JANUARY TO DECEMBER 1945 · TO THE WAR'S END

If No. 452 Squadron, or indeed if any in the RAAF higher command thought that they would be involved in MacArthur's push towards Japan, they would be very disappointed. MacArthur as a true American Ceasar, wanted all the laurels of defeating Japan and so it would be Americans who would force the Japanese surrender. Australia, the British and the Dutch would be left to mop up.

As the war entered its sixth year, the Australian Spitfire squadrons now came under the command of the 1st Tactical Air Force, an organisation formed in October 1944 that was meant to be mobile and flexible and able to take the fight to the enemy, unhindered by the geographic boundaries of previous years. At that time, the 1st TAF comprised five wings: one attack wing; two fighter wings and two airfield construction wings. Assigned under their respective wings, were squadrons of Beaufighters, Spitfires, and Kittyhawks. On joining the 1st TAF, Nos 452 and 457 Squadrons were combined into No. 80 Wing in December 1944 with a newcomer from New Guinea also flying Spitfires, No. 79 Squadron.

For No. 452 Squadron, January 1945 began with standing night patrols over the Halmaheras and strafing sorties against Japanese coastal barges and villages. These attacks were not without danger. On 7 January, Pilot Officer Hugh MacNeil's aircraft was hit by ground fire and unable to return to base, he crashed into trees at Cape Gila. With bad head injuries, he too would not return to the squadron after hospitalisation.

A week later, No. 452 Squadron suffered three more losses. Flight Sergeant Ed Stevenson suddenly disappeared over Ternate Island and although searchers found his aircraft slightly

The No. 452 Squadron camp at Morotai.

submerged just off the coast, no trace of Stevenson was found. It transpired that Stevenson had been captured by the Japanese becoming a prisoner of war. He died of malaria in captivity three months later. Then just five days after Stevenson went missing, another crash took the life of Flight Lieutenant Les Leeming who suffered a night engine failure. On attempting to land, he struck a bank short of the runway which killed him instantly. The third loss was Leading Aircraftman Ron Orpen. Orpen had for some reason tossed a lit match into what was supposed to be an empty Spitfire belly tank. The tank exploded killing him instantly.

The first enemy raid for two months arrived over Morotai on the night of 22 March with intruders dropping bombs to the north before escaping the ground fire sent up to stop them. There was little damage. Then, drama. In April, Air Commodore Harry Cobby, the commander of the 1st TAF, received eight identically worded resignations from high-ranking squadron and wing personnel. Their assertions were that they were flying dangerous operations against what they considered "senseless unimportant ground targets". This became known as the "Morotai Mutiny", with the result that Cobby and two group captains were relieved of their commands. Caldwell, one of the group captains, was already under charges facing a court-martial in that he had exchanged whisky (which he had purchased) with the Americans in exchange for building materials to improve the living conditions for the men. This was against orders. Whether it was jealousy or the ignorance of local conditions by those down south the charges stood, and Caldwell was demoted. The affair resulted in a legal inquiry under Justice Barry, but that is another story.

At squadron level life continued and so did the war. Most flights now involved intruder sweeps or air alerts and none engaged the enemy in the air. By now, the IJA and IJN air forces had either been destroyed or recalled to the Philippines. With little air threat it was No. 452 Squadron's turn to become dive-bombers for the Borneo campaign. By early May 1945, the Allies were preparing to invade Borneo. A series of operations known as *Oboe* were planned to sequentially take the island. The first, *Oboe 1*, was to capture the small but strategically

Spitfire Mk VIII A58-524 (QY-C) which only spent a month with No. 452 Squadron before it crashed on Morotai on 18 January 1945. The pilot, Flight Lieutenant Les Leeming, was killed.

Bomb racks fitted to a Spitfire's underwing pylons. With little aerial opposition in the final months of the war No. 452 Squadron's Spitfires were utilised as dive-bombers.

significant island of Tarakan. The amphibious assault commenced on 1 May at which time No. 452 Squadron was warned of a move to the Japanese airstrip on Tarakan which was captured five days later. Then followed Balikpapan, the Netherlands East Indies' largest and most important oil refinery and port. The refinery provided the Japanese the bulk of their lubricating oils and aviation fuel needed to sustain their war effort. So important was aviation fuel to the Japanese war machine, that ceasing the supply of it has been credited as one of the prime reasons for Japan's entry into the war.

With much haste, the bulk of No. 452 Squadron's personnel and stores were embarked on HMAS *Westralia* and *Manoora* on 4 May and arrived at Tarakan three days later. Here the main work was to repair Juwata airstrip which required perforated steel planking to be laid in preparation for the Spitfires, and to clean up the mess left by the enemy. This took to the end of June when the airstrip was officially named Croydon. The airstrip was now 1,600 yards long with steel matting laid over the mud, and with fifty-foot aprons on either side. Once ready, the Spitfires began to arrive from 29 May with all in place ten days later.

Bomb racks were fitted under the wings and training commenced but inadvertent releases and bomb hang-ups plagued the squadron's operations. These problems were soon solved when the army captured a Japanese bomb dump, and the squadron's armourers found the Japanese bombs to be better suited to the Spitfires. According to pilot Ted Sly:

> ... using the Japanese 250-lb bombs we developed a fairly accurate method of dive-bombing bridges with the Spitfires.

Sly recalled that the squadron was later castigated by a visiting RAAF "chairborne" wing commander from down south for making the unapproved modification.

Still Spitfire accidents continued. In almost a repeat of Leeming's accident earlier in the year, on 22 May Flying Officer Clive Miller was killed at Morotai when he undershot on landing. He crashed into a bank of coral at the end of the strip and the aircraft burst into flames. On 5 June Squadron Leader Kel Barclay took over as commanding officer from Squadron Leader Birch. As if to welcome him to the action, Barclay was hit by anti-aircraft fire on 12 July and baled out. His aircraft A58-647 (QY-V) was lost but Barclay in his dinghy was picked up by a rescue launch three hours later. Ten days earlier, on 2 July, Flight Lieutenant Alan Proctor was shot down and crashed into a hill at Sembackoeng, Borneo. Unfortunately, he could not escape the blazing wreckage. He was not the only loss that month. On 10 July, Flight Lieutenant Norm Cullen was killed when his Spitfire A58-518 (QY-O) hit the trees while pulling out of a bombing run over Tawao in North Borneo.

To support the Allied invasion, on 15 July a detachment of eleven Spitfires led by Barclay flew south to Balikpapan on the central east coast of Borneo. Here they operated from Sepinggan strip, a 1,000-yard single runway prepared by the airfield construction squadron and an operational base unit. Two days later, No. 452 Squadron flew its first sorties of close air support for Australian troops. The Spitfires strafed Japanese entrenchments. Yet still accidents continued. Upon returning, Flight Sergeant Brian Gurney misjudged his landing and touched down short. The soft mud caused the aircraft to cartwheel and nose over. It then burst into flames. Gurney was lucky to escape with a fright, but the Spitfire A58-516 (QY-T) was destroyed. Armourer, Leading Aircraftman Ken Gordon, was not so lucky. The next day, while pumping up a fuel burner, he was killed when the burner exploded.

Despite the tragic loss of so many airmen, operations out of Balikpapan were certainly a morale booster. Finally, some action occurred. On 19 July, the unit historian was able to report:

> The most effective operation against the enemy for two years was thoroughly enjoyed by pilots on barge sweeps among the numerous waterways and rivers of upper Balikpapan Bay. Fifteen sorties were flown and the score at the end of the day was eleven barges burnt or sunk and three damaged to the extent that they might well be claimed probably destroyed.
>
> Probably the highlight of the day was provided by Flight Lieutenant Blumer when he attacked a barge on which [there were enemy troops]. The cargo carried must be a matter for speculation as the first burst, fired from short range, caused an explosion with smoke and flame through which Flight Lieutenant Blumer passed. Proof of the incident was provided by the film of oily soot that covered the aircraft, the scorched paint on the airscrew and the singeing of the pilot's arm, though the cockpit hood was closed.

That this operation was claimed as the most effective for two years confirms the previous complaints made by Caldwell and his pilots.

The last air combat of No. 452 Squadron's war and the first for six months was on 24 July

Spitfire Mk VIII A58-516 (QY-T). This fighter was received by No. 452 Squadron in December 1944 and was regularly flown by Flight Lieutenant Lyn Crompton. It was destroyed in a landing accident at Balikpapan on 17 July 1945.

A fine photograph of Spitfire Mk VIII A58-516 (QY-T) which is illustrated in the profile above.

1945 when Flying Officer Jeff "Argus" King made an early morning contact with a flight of ten enemy aircraft. He made two attacks and shot down a Nakajima Ki-49 Helen bomber into the sea in flames. No. 452 Squadron's last operational sortie of the war was flown along the Milford Highway on 10 August. Fittingly led by Squadron Leader Barclay, one truck was destroyed and four were left badly damaged.

On 15 August, news arrived of the Japanese surrender which was met "without excitement or demonstration". So ended No. 452 Squadron's war. All offensive operations were suspended although night readiness was still maintained. As if to put another damper on squadron morale, Flying Officer Ed Lee died eleven days after his Spitfire A58-650 collided with a truck parked at the edge of the runway just after he became airborne.

With the war at a close, several changes were made to both No. 452 Squadron's command and the flying routine. Replacing Squadron Leader Barclay, on 12 September Flight Lieutenant William Friend assumed command as temporary commanding officer and was given the task by wing headquarters of preparing the Spitfires to return to Australia. Friend had flown

Spitfires over Europe and the Middle East with the RAF and was no doubt pleased to be back with his countrymen.

On 27 October, a new commanding officer briefly replaced Friend who returned to Australia for discharge, taking a Spitfire with him. Transferring from No. 79 Squadron, Flight Lieutenant Len Reid would only spend five days in Tarakan before flying the squadron home. With his departure, left to tidy things up was the unit administrative officer, Flying Officer Ken Williams, who held the position of temporary commanding officer until the unit officially disbanded on 17 November 1945.

As for the Spitfires, they would end up at No. 6 Aircraft Depot at Oakey in Queensland awaiting their cutting up and scrapping. It was a sad end to such a graceful aircraft.

Although the war was well over, the No. 452 Squadron Roll of Honour (see Appendix 5) contains one further casualty in 1946. Leading Aircraftman Lewis Percival was one of the squadron's cooks who had joined the unit in December 1942. He was discharged as medically unfit for further service in January 1945 and died on 26 February 1946. His death in Brisbane was later determined to be caused by his war service.

Recent events have shown that 1945 was not the end of No. 452 Squadron. The squadron's rebirth resulted from the high level of administrative control that had been placed upon No. 44 Wing, a new wing raised in December 2000 to consolidate air traffic control (ATC) across the air force. No. 44 Wing was a unit under the RAAF's Surveillance and Response Group. The eleven geographically dispersed ATC detachments all reporting to the one organisation became unmanageable, so the RAAF decided to create two new units to distribute the workload. These were Nos. 452 and 453 Squadrons, both WWII squadrons that had served with distinction.

By Air Force Organisation Directive, No. 452 Squadron officially reformed on 1 December 2010. The squadron became responsible for ATC services for both military and civilian air operations around RAAF Bases Darwin, Tindal, Townsville and Amberley and the Army Aviation Centre at Oakey. No. 452 Squadron now lives on.

Spitfires awaiting their fate in the No. 6 Aircraft Depot "graveyard" at Oakey in Queensland. These aircraft including QY-X were scrapped between 1945 and 1948.

APPENDIX 1

NO. 452 SQUADRON COMMANDING OFFICERS

Name	From	To	Remarks
Formed Kirton-in-Lindsey	8 Apr 1941		
SQNLDR Roy Gilbert Dutton	13 Apr 1941	12 Jun 1941	
SQNLDR Robert Wilton Bungey	12 Jun 1941	25 Jan 1942	
SQNLDR Keith William Truscott	25 Jan 1942	18 Mar 1942	
SQNLDR Raymond Edward Thorold-Smith	18 Mar 1942	15 Mar 1943	KIA
SQNLDR Clive Robertson Caldwell	16 Mar 1943	29 Mar 1943	Temporary CO
SQNLDR Ronald Sommerville MacDonald	30 Mar 1943	3 Feb 1944	
SQNLDR Louis Thomas Spence	3 Feb 1944	27 Nov 1944	
SQNLDR Arthur Hugh Birch	27 Nov 1944	4 Jun 1945	
SQNLDR Kelvin Milne Barclay	5 Jun 1945	12 Sep 1945	
FLTLT William Friend	12 Sep 1945	27 Oct 1945	Temporary/CO
FLTLT Leonard Stanley Reid	27 Oct 1945	31 Oct 1945	
FLGOFF Kenneth John Williams	31 Oct 1945	17 Nov 1945	Temporary/CO
Disbanded	17 Nov 1945	-	Disbanded
Reformed	-	1 Dec 2000	
Wing Commander Robert Morris	1 Dec 2010	-	Reformed

APPENDIX 2

NO 452 SQUADRON HONOURS & AWARDS

Europe

Name	Award	Date and Gazettal
Flying Officer Brendon Fergus Finucane[1]	DFC	LG 25 April 1941 p. 2745
Flying Officer Brendon Fergus Finucane	Bar to DFC	LG 22 August 1941 p. 5217
Flying Officer Brendon Fergus Finucane	Second Bar to DFC	LG 15 Sep 1941 p. 5595
Squadron Leader Robert Wilton Bungey	DFC	LG 7 October 1941 p. 5773
Flying Officer Brendon Fergus Finucane	DSO	LG 11 October 1941 p. 6103
Squadron Leader Keith William Truscott	DFC	LG 17 October 1941 p. 6035
Sergeant Keith Bruce Chisholm	DFM	LG 17 October 1941 p. 6035
Flight Lieutenant Raymond Edward Thorold-Smith	DFC	LG 2 December 1941 p. 6887
Squadron Leader Keith William Truscott	Bar to DFC	LG 27 March 1942 p. 1381
Flying Officer Clive Newton Wawn	DFC	LG 10 April 1942 p. 1593
Squadron Leader Keith William Truscott	MiD[2]	LG 1 April 1943 p. 1414
Flight Lieutenant Keith Bruce Chisholm	MC	LG 13 February 1945 p. 882
Flight Lieutenant Keith Bruce Chisholm	Polish Cross with Swords	
Flight Lieutenant Desmond John Cormack	DFC	LG 22 February 1946 p. 1075
Flying Officer Jeffrey Cecil King	MiD	LG 25 June 1946 p. 3218
Flight Lieutenant Timothey O'Connell	MiD	LG 25 June 1946 p. 3218.
Flying Officer Jack Emerson Pretty	DFC	LG 27 June 1946 p. 3218
Warrant Officer Clive Leslie Swift	MiD	LG 27 June 1946 p. 3219

Pacific

Name	Award	Date and Gazettal
Squadron Leader Ronald Sommerville Macdonald[3]	MiD	

Notes:

Armstrong, Barclay, Brennan, Goldsmith, Bisley, Spence and Cundy awarded a DFC but not when in No. 452 Squadron.

DFC	Distinguished Flying Cross	LG	London Gazette
DFM	Distinguished Flying Medal	MC	Military Cross
DSO	Distinguished Service Order	MiD	Mentioned in Despatches

1 Awarded while at No. 452 Sqn.
2 Not with No. 452 Sqn
3 For No. 12 Sqn.

APPENDIX 3

NO. 452 SQUADRON ACES & CLAIMS

Europe

Aces	Claims Self + shared, probable, damaged	Remarks
Keith Bruce Chisholm MC, Polish Gold Cross with Swords	5+2, 0, 1	All with No. 452 Sqn
Brendan Eamonn Fergus Finucane DSO, DFC**	26+6, 8+1, 8	17+2, 2, 2 with No. 452 Sqn
Andrew Henry Humphrey	7, 2, 0	1 with No. 452 Sqn
Raymond Edward Thorold-Smith DFC	6+1, 0, 1	All with No. 452 Sqn
Keith William Truscott DFC*	14, 3, 3	All bar one destroyed with No. 452 Sqn

Europe Other Claims

Pilot	Claims	Remarks
Robert Wilton Bungey (RAF) DFC	1+2, 0, 0	One with No. 452 Sqn
William Oswald Cable DFC	3, 0, 3	None with No. 452 Sqn
Raife James Cowan	1, 0, 0	On 2 Oct 1941
Alfred Graham Douglas	3, 1, 2	2, 0, 0 with No. 452 Sqn
Bruce Paul Dunstan (RAF)	1, 0, 0	On 20 Sep 1941
William Davies Eccleton	1, 0, 0	On 3 Aug 1941
John Marsh Emery	1, 1, 0	On 13 Oct 1941
Ray Harold Charles Sly	2, 1, 2+1	2 ,0, 0 with No. 452 Sqn
Archibald Roger Stuart	3, 0, 0	All with No. 452 Sqn
Eric Blair Tainton	1, 0, 0	On 16 Aug 41
Clive Newton "Bardie" Wawn DFC	3, 2, 3	2, 1, 2 with No. 452 Sqn
William Douglas Willis	1, 0, 0	On 2 Sep 1941

Australia

Aces	Claims	Remarks
Hugo Throssell Armstrong DFC*	10+1, 3, 2	None with No. 452 Sqn
John Henry Eric Bisley DFC	6+1, 1, 0	1 with No. 452 Sqn
Virgil Paul Brennan DFC, DFM	10, 1, 6	None with No. 452 Sqn
William Ronald Cundy DFC, DFM	5+2, 1, 1	None with No. 452 Sqn
Adrian Philip Goldsmith DFC, DFM	16+1, 2, 7	4, 0, 1 with No. 452 Sqn

Australia Other Claims

Pilot	Claims	Remarks
Edward Smith Hall	2+1 shared, 1, 2	2,0,2 with No. 452 Sqn
Allen Mawer	3, 0, 2	2, 0, 1 with No. 452 Sqn
Ronald Somerville MacDonald	2, 0, 1	All with No. 452 Sqn
Davison Freeman Evans	1, 1, 0	All with No. 452 Sqn

Notes:

Claims are listed as: d, p, d - destroyed, probably destroyed, damaged; "+" indicates shared with another pilot.

e.g. Thorold-Smith's 6+1, 0, 1 indicated six destroyed plus one shared destroyed, none probably destroyed and one damaged.

APPENDIX 4

NO. 452 SQUADRON TALLY BOARD & LOSSES

Europe - 1941

Date	Claim	Pilot	Total	Loss	Pilot	Remarks
5 Jul 1941	-	-	-	Spitfire IIa P8085	Costello	KIA
11 Jul 1941 Sweep	Bf 109F	Finucane	1, 0, 0	Spitfire IIa P7562	Roberts	POW, Escaper
24 Jul 1941 Sweep Cherbourg	Bf 109E	Humphrey	1, 0, 0	-	-	-
3 Aug 1941 Sweep	Bf 109F Bf 109F (p) Bf 109F	Finucane Finucane Eccleton	2, 1, 0	-	-	-
9 Aug 1941 Circus 68 Gosnay	Bf 109E Bf 109F (½) Bf 109F (½) Bf 109E Bf 109F (½) Bf 109F (½) Bf 109F (½) Bf 109F (½)	Finucane Finucane Finucane Truscott Lewis Th-Smith Chisholm Chisholm	5, 0, 0	Spitfire IIa P7682 Spitfire IIa P7590 Spitfire IIa P8361	O'Byrne Chapman Haydon	POW Stalag L3 KIA KIA
16 Aug 1941 Circus 74 Marquise	Bf 109E	Finucane	1, 0, 0	-	-	First Wave
Circus 75 St Omer	Bf 109F Bf 109F Bf 109E Bf 109E Bf 109E Bf 109E Bf 109E	Finucane Finucane Truscott Chisholm Chisholm Tainton Stuart	7, 0, 0	-	-	Second Wave
19 Aug 1941 Circus 81 Gosnay	Bf 109F Bf 109F (p) Bf 109F (p)	Finucane Finucane Truscott	1, 2, 0	Spitfire Vb P8717 Spitfire Vb AB785 Spitfire Vb AB794	Eccleton Gazzard Willis	KIA KIA WIA - OK

Date	Claim	Pilot	Total	Loss	Pilot	Remarks
26 Aug 1941 Circus 87 St Omer	Bf 109E Bf 109E	Douglas Stuart	2, 0, 0	-	-	-
27 Aug 1941 Circus 85 St Omer	Bf 109E Bf 109E Bf 109E Bf 109E	Finucane Finucane Th-Smith Th-Smith	4, 0, 0	-	-	-
2 Sep 1941 Roadstead	Bf 109E Bf 109E	Willis Stuart	2, 0, 0	-	-	-
7 Sep 1941 Practice attacks	-	-	-	Spitfire Vb AB874 Spitfire Vb P8716	Hanigan Williams	KIFA (mid-air collision) KIFA (mid-air collision)
18 Sep 1941 Circus 99 Rouen	Bf 109E Bf 109E Bf 109F Bf 109F Bf 109F (p) Bf 109F (d) Bf 109F (d)	Th-Smith Chisholm Douglas Truscott Truscott Wawn Wawn	4, 1, 2	Spitfire Vb W3512 Spitfire Vb P8703 Spitfire Vb W3508 Spitfire Vb W3600	Willis Stuart Try Manning	KIA POW Stalag L7 POW Stalag L4 KIA
20 Sep 1941 Circus 100B Abbeville	Bf 109F Bf 109F Bf 109F Bf 109F Bf 109E Bf 109F (d) Bf 109F Bf 109E Bf 109F (d) Bf 109F (d)	Finucane Finucane Finucane Truscott Truscott Truscott Chisholm Dunstan Elphick Elphick	7, 0, 3	Spitfire Vb AB841	Milne	POW Stalag 357
21 Sep 1941 Circus 101 Gosnay	Bf 109F Bf 109F Bf 109E Bf 109F Bf 109F Bf 109F (d) Bf 109 (d)	Finucane Finucane Truscott Wawn Chisholm Chisholm Lewis	5, 0, 2	-	-	-
2 Oct 1941 Circus 104 Le Havre	Bf 109F Bf 109F (d) Bf 109F	Finucane Finucane Cowan	2, 0, 1	-	-	-

Date	Claim	Pilot	Total	Loss	Pilot	Remarks
12 Oct 1941 Circus 107 Boulogne	Bf 109F Bf 109F (d) Bf 109F (p)	Finucane Truscott Truscott	1, 1, 1	Spitfire Vb W3520	Chisholm	POW Stalag VIIIB, escaper
13 Oct 1941 Circus 108A Arques	Bf 109F Bf 109F Bf 109F (d) Bf 109F Bf 109F (d) Bf 109F Bf 109F Bf 109F Bf 109F (d) Bf 109F	Finucane Finucane Finucane Th-Smith Th-Smith Truscott Truscott Emery Emery Schrader[1]	7, 0, 3	Spitfire Vb AB852 Spitfire Vb AD310	Jackson Elphick	KIA Abandoned, Rescued
6 Nov 1941 Escort Cap Gris Nez	Bf 109F FW 190 Bf 109F (d) Bf 109F Bf 109 (d) Bf 109F (d)	Truscott Th-Smith Th-Smith Bungey Lewis Sly	3, 0, 3	Spitfire Vb AD430 Spitfire Vb AD242	Geissmann Schrader	KIA KIA
8 Nov 1941 Circus 110 Lille	Bf 109F Bf 109F	Truscott Truscott	2, 0, 0	Spitfire Vb AB842 Spitfire IIa P8645	Truscott Dunstan	Ditched, Rescued Ditched, Rescued
8 Dec 1941 Escort	-	-	-	Spitfire Vb AB966	Emery	KIA

End of 1941 - Total 57 destroyed, 5 probably destroyed and 15 damaged.

Europe – 1942

Date	Claim	Pilot	Total	Loss	Pilot	Remarks
4 Jan 1942 Cannon test	-	-	-	Spitfire Vb AB195	Baker	Ditched, drowned
22 Jan 1942 Minesweeper protection	-	-	-	Spitfire AB992	Lewis	Ditched, drowned
13 Feb 1942 Rodeo Sweep	He 114 (s)	-	1, 0, 0	-	-	Shared with numerous others. Not officially credited to any
15 Feb 1942 Patrol Scramble	Bf 109F (d)	Makin	0, 0, 1	Spitfire Vb P8711	Harper	KIA

1 Schrader claimed a probable which was later upgraded.

Date	Claim	Pilot	Total
9 Mar 1942 Circus 113 Marzingarbe	Bf 109F Bf 109F (d) Bf 109F Bf 109F (p)	Truscott Truscott Wawn Wawn	2, 1, 1
13 Mar 1942 Circus 114 Hazebrouck	FW 190	Sly	1, 0, 0
14 Mar 1942 Ramrod 15 Le Havre	FW 190 Bf 109F (p)	Truscott Morrison	1, 1, 0
10 Apr 1942 Formation	-	-	-
3 May 1942 Patrol	Ju 88 (d)	Lamerton	0, 0, 1
7 May 42	-	-	-
8 May 1942 Practice gunnery	-	-	-

Loss	Pilot	Remarks
Spitfire Vb AA849	Hamilton	KIA
-	-	-
-	-	-
Spitfire Vb BL384	Waters	Ditched, drowned
Spitfire Vb AD537	Lamerton	Abandoned, rescued
Spitfire Vb AB171	Whillans	FTL, safe
Spitfire Vb BL351 Spitfire Vb AB244	Goodhew Ford	KIFA (mid-air) Inj, Rescued

End of 1942 - 4+1 He114 destroyed, 2 probably destroyed, 3 damaged

End for Europe - 62 destroyed, 7 probably destroyed, and 18 damaged

19 Pilots Killed, 6 PoW, 2 Escapers

Notes:

(d) damaged (p) probable (s) shared

KIA – Killed in action KIFA – Killed in Flying Accident POW – Prisoner of War WIA – Wounded in action

No. 452 Squadron Tally Board & Losses

Pacific – 1942

Date	Claim	Pilot	Total
9 Oct 42	-	-	-

Loss	Pilot	Remarks
Spitfire Vc BR471/A58-25	Clifford	KIFA (Crashed into sea)

Pacific – 1943

Date	Claim	Pilot	Total
27 Jan 43	-	-	-
27 Feb 43	-	-	-
7 Mar 43	-	-	-
15 Mar 43 Raid 53	Betty Hap Hamp (d) Hap (d)	Goldsmith Goldsmith Goldsmith Hall	2, 2, 0
2 May 43 Raid 54	Betty Zeke Zeke (d) Zeke (p) Zeke (d)	Goldsmith Mawer Mawer Evans Hall	2, 1, 2
20 Jun 43 Raid 55	Helen Helen Helen Zeke Helen (d) Helen (d) Zeke (d)	MacDonald Bisley Evans Mawer Whillans Cross Colyer	4,0,3
23 Jun 43	-	-	-
28 Jun 43 Raid 56	-	-	-

Loss	Pilot	Remarks
Spitfire Vc BS184/A58-73	Hutchinson	KIFA (mid-air collision with Stockley in BR549/ A58-55
Spitfire Vc BS175/A58-69	Ford	KIFA (crashed enroute to WA)
Spitfire Vc BS237/A58-98	Moore	KIFA (Eng failure)
Spitfire Vc BS231/A58-92	Thorold-Smith	KIA
Spitfire Vc BS293/A58-101	Lloyd	Abandoned, pilot safe
Spitfire Vc BS162/A58-12	McNab	KIA
Spitfire Vc BR547/A58-53	Stagg	Baled out, rescued
Spitfire Vc BS226/A58-90	Nichterlein	Belly landed
Spitfire Vc BS225/A58-89	Fox	Abandoned, pilot safe
Spitfire Vc BR526/A58-34	Goldsmith	Shot down, rescued
Spitfire Vc EE607/A58-107	Nichterlein	KIA
Spitfire Vc BS174/A58-68	Ruskin-Rowe	KIA
Spitfire Vc AR510/A58-1	Williams	Engine fail, baled out, safe
Spitfire Vc EE608/A58-108	Cowell	FTL on beach, rescued

Date	Claim	Pilot	Total
30 Jun 43 Raid 57	Zeke Betty (½)[2] Betty (d) Betty (d)[3]	Hall Tully Lloyd Cowell	1½, 0, 1½
6 Jul 43 Raid 58	Betty Zeke Zeke (d)	MacDonald Lloyd Tully	2, 0, 1
10 Aug 43	Pete Pete (½) (p) Pete (½) (p)	Young Young Coombes	1, 1, 0
13 Aug 43 Raid 59	-	-	-
14 Aug 43 Raid 60	-	-	-
21 Aug 43 Raid 61	-	-	-
7 Sep 43 Escorted Recce	Tony Oscar (d) Oscar (p) Zeke (d) Oscar (d)	Goldsmith Bassett Cowell Coombes King	1, 1, 3
15 Sep 43 Raid 62	-	-	-
18 Sep 43 Raid 63	-	-	-
26 Sep 43	-	-	-
12 Nov 43 Raid 64	-	-	-

Loss	Pilot	Remarks
Spitfire Vc BR241/A58-19 Spitfire Vc AR523/A52-2 Spitfire Vc BR546/A58-52	Lamerton Duncan Cross	KIFA (glycol leak) Engine fail, rescued FTL, rescued
Spitfire Vc BR497/A58-32 Spitfire Vc BS193/A58-79 Spitfire Vc BR237/A58-15	Lloyd Makin Richardson	Shot down, pilot safe Engine fail, safe Engine fail, baled out, safe
-	-	-
-	-	-
-	-	-
-	-	-
Spitfire Vc LZ884/A58-227 Spitfire Vc BR549/A58-55	MacDonald Tully	Shot down, inj Abandoned, pilot safe
-	-	-
-	-	-
Spitfire Vc EE677/A58-121 Spitfire Vc JL314/A58-201	Adam Mawer	KIFA (mid-air collision) KIFA (mid-air collision)
-	-	-

Zeke also listed as Hap and Hamp.

End of 1943 - 13½ destroyed, 5 probably destroyed, 10½ damaged

2 Shared with SQNLDR Eric Gibbs (RAF), No. 54 Sqn.

3 Shared with SQNLDR Eric Gibbs (RAF), No. 54 Sqn who shot it down.

Pacific - 1944

Date	Claim	Pilot	Total
24 Apr 44	-	-	-
12 Jun 44	Dinah	Gamble (⅓) Beaton (⅓) O'Loughlin (⅓)	1, 0, 0
18 Sep 44	-	-	-
17 Dec 44	-	-	-
24 Dec 44	Nick	Pretty	1, 0, 0
30 Dec 44	Oscar	May	0,0,1

Loss	Pilot	Remarks
Spitfire Vc MA353/A58-232	Dunning	KIFA
Spitfire Mk Vc EE673/A58-117	-	O'Loughlan engine failure baled out OK
Spitfire VIII JG622/A58-435	Kelly	KIFA x 2 (Collided with B-24)
Spitfire VIII MT542/A58-478	Sturm	KIFA x 2 (ground collision)
Spitfire VIII MT620/A58-519	Markwell	KIFA (EFATO – ditched)
-	-	-

End of 1944 – 2 Destroyed, 0 probably destroyed, 1 damaged

Pacific - 1945

Date	Claim	Pilot	Total
1 Jan 45	-	-	-
18 Jan 45	-	-	-
22 May 45	-	-	-
2 Jul 45	-	-	-
10 Jul 45	-	-	-
24 Jul 45	Helen	King	1, 0, 0

Loss	Pilot	Remarks
Spitfire VIII JG549/A58-419	Stevenson	Shot down, POW, died of Malaria 14 Apr 45
Spitfire VIII MT549/A58-524	Leeming	KIFA (crashed on landing)
Spitfire VIII MT778/A58-533	Miller	KIA
Spitfire VIII MT545/A58-465	Proctor	KIA
Spitfire VIII MT618/A58-518	Cullen	KIA
-	-	-

End of 1945 – 1 destroyed, 0 probably destroyed, 0 damaged

Total for Pacific - 16½ destroyed, 5 probably destroyed, 11½ damaged

18 Pilots killed, 1 PoW

Total Claims - 78½ destroyed, 12 probably destroyed, 29½ damaged

Notes:

(d) damaged (p) probable (s) shared

KIA – Killed in action KIFA – Killed in Flying Accident POW – Prisoner of War WIA – Wounded in action

Sources:

No. 452 Squadron ORB and Misc Files (includes Combat Claims) – Office of Air Force History, Canberra.

Foreman, John, *Fighter Command War Diaries, Vol 2, September 1940 to December 1941*, Air Research Publications, Walton-on-Thames, 1998.

Foreman, John, RAF *Fighter Command Victory Claims, Part 2, 1 January 1941 to 30 June 1943*, Red Kite, Walton-on-Thames, 2005.

Franks, Norman L. R., *Royal Air Force Fighter Command Losses of the Second World War 1939-1941*, Midland Publishing, Leicester, 1997.

Morgan, Eric B. and Shacklady, Edward, *Spitfire: The History*, Key Books, Stamford, 1987.

Prisoners of War Naval and Air Forces of Great Britain and the Empire 1939-1945, J.B. Hayward and Son, Polstead.

Shores, Christopher and Williams, Clive, *Aces High*, Grub Street, London, 1994.

Shores, Christopher, *Aces High Vol 2*, Grub Street, London, 1999.

Shores, Christopher, *Those Other Eagles*, Grub Street, London, 2004.

Storr, Alan, *RAAF Fatalities in Second World War in RAAF Squadrons and Support Units*, AWM.

Tom Roberts, *Wingless: A Biographical Index of Australian Airmen Detained in Wartime*, Self-published, Ballarat,2011.

www.airhstory.org.uk/spitfire/home.html

APPENDIX 5

NO. 452 SQUADRON ROLL OF HONOUR

Europe

Name	Date/Aircraft	Buried
Sergeant Andrew Gordon Costello RAAF No. 404086. Age 23	5 July 1941 IIa P8085	Block A. Grave 190. Kirton-in-Lindsey Burial Ground
Sergeant Christopher Geoffrey Blomfield Chapman RAAF No. 404198. Age 28	9 Aug 1941 IIa P7590	Longuenesse (St Omer) Souvenir Cemetery, Plot 9, Row C, Grave 21
Sergeant Gerald Barrington Haydon RAAF No. 404100. Age 19	9 Aug 1941 IIb P8361	Longuenesse (St Omer) Souvenir Cemetery, Plot 9, Row C, Grave 20
Flying Officer William Davies Eccleton RAAF 402232. Age 25	19 Aug 1941 Vb P8717	Dunkirk Town Cemetery, Plot 2, Row 2, Grave 14
Sergeant Richard George Gazzard RAAF No. 402115. Age 21	19 Aug 1941 Vb AB785	Oostduinkerke Communal Cemetery, Row G, Grave 167
Sergeant James Neate Hanigan RAAF No. 402120. Age 24	7 Sep 1941 Vb AB874	Whyteleafe (St Luke) Cemetery Row H, Grave 32
Sergeant Kenneth Victor Williams RAAF No. 400264. Age 25	7 Sep 1941 Vb P8716	Whyteleafe (St Luke) Cemetery Row H, Grave 33
Flying Officer William Douglas Willis RAAF No. 400166. Age 24	18 Sep 1941 Vb W3512	Ste. Marie Cemetery, Le Havre, Div 67, Row D, Grave 12
Sergeant Charles Frederick Richard Manning RAAF No. 400230. Age 20	18 Sep 1941 Vb W3600	St Aubin-Sur-Mer Churchyard, France
Sergeant Edgar Purton Jackson RAAF No. 400227. Age 21	13 Oct 1941 Vb AB852	Runnymede Memorial, Panel 62
Sergeant Bernard Malcolm Geissmann RAAF No. 404334. Age 19	6 Nov 1941 Vb AD430	Runnymede Memorial, Panel 62
Sergeant Eric Henry Schrader RAAF No. 400135. Age 21	6 Nov 1941 Vb AD242	Runnymede Memorial, Panel 63
Sergeant John Marsh Emery RAAF No. 407116. Age 27	8 Dec 1941 Vb AB966	Runnymede Memorial, Panel 62
Pilot Officer Allenby Leslie Baker RAAF No. 403159. Age 23	4 Jan 1942 Vb AB195	Runnymede Memorial, Panel 111
Flying Officer Donald Edwin Lewis RAAF No. 400148. Age 19	22 Jan 1942 Vb AB992	Runnymede Memorial, Panel 110
Sergeant Francis Gilbert Harper RAAF No. 404664. Age 28	15 Feb 1942 Vb P8711	Runnymede Memorial, Panel 112
Sergeant Malcolm Wallace Hamilton RAAF No. 404726. Age 24	9 Mar 1942 Vb AA849	Runnymede Memorial, Panel 112

Name	Date/Aircraft	Buried
Sergeant Ralph Ernest Waters RAAF No. 408165. Age 25	10 Apr 1942 Vb BL384	Runnymede Memorial, Panel 113
Sergeant Reginald Allan Goodhew RAAF No. 403332. Age 23	8 May 1942 Vb BL351	Andreas (St Andrew) Churchyard, Service Plot Grave 5

Pacific

Name	Date	Buried
Sergeant Michael Clifford RAAF No. 403320. Age 26	9 Oct 1942 Vb BR471 A58-25	Botany General Cemetery R.C. Plot Sec 3, Grave 71.
Flight Lieutenant John Robertson Ross RAAF No. 400157. Age 26	20 Oct 1942 Ryan A50-21	Brighton General Cemetery C of E Plot Compt. G, Grave 61.
Leading Aircraftman Walter Clarence Elliott RAAF No. 14856. Age 29.	21 Oct 1942 Ryan A50-21	Sydney War Cemetery Grave A.A.13
Sergeant Eric Ebsworth Hutchinson RAAF No. 403602 Age 26.	27 Jan 1943 Vc BS184 A58-73	Adelaide River War Cemetery R.C.10
Flying Officer William Hendrie Ford RAAF No. 401429 Age 24.	27 Feb 1943 Vc BS175 A58-69	Adelaide River War Cemetery Grave R.D.7
Flight Sergeant Eric Mervyn Moore RAAF No. 403357 Age 22	7 Mar 1943 Vc BS237 A58-98	Adelaide River War Cemetery Grave R.D.1
Squadron Leader Raymond Edward Thorold-Smith RAAF No. 402144 Age 24	15 Mar 1943 Vc BS231 A58-92	Northern Territory Memorial Panel 6
Flying Officer Alexander Charles McNab RAAF 405420 Age 24	2 May 1943 Vc BS162 A58-12	Northern Territory Memorial Panel 7
Pilot Officer Willie Everard Nichterlein RAAF No. 416104 Age 23	20 Jun 1943 Vc EE607 A58-107	Northern Territory Memorial Panel 7
Pilot Officer Anthony Thomas Ruskin-Rowe RAAF No. 411389 Age 24	20 Jun 1943 Vc BS174 A58-68	Adelaide River War Cemetery Grave F.A.7
Flying Officer William John Lamerton RAAF No. 407900 Age 23	30 Jun 1943 Vc BR241 A58-19	Adelaide River War Cemetery Grave F.A.4
Flight Lieutenant Albert Ernest Watkin RAAF No. 405265 Age 27	16 Nov 1943 US C-47A	Rockhampton War Cemetery Plot A, Row A, Grave 8

Name	Date	Buried
Leading Aircraftman Stanley Kirk Sims RAAF No. 28071 Age 22	16 Nov 1943 US C-47A	Rockhampton War Cemetery Plot A, Row A, Grave 11
Flying Officer John Philip Adam RAAF No. 404491. Age 30	26 Sep 1943 Vc EE677 A58-121	Adelaide River War Cemetery Grave F.C.8
Flying Officer Granville Allen Mawer RAAF No. 403112. Age 23	26 Sep 1943 Vc JL314 A58-201	Adelaide River War Cemetery Grave F.C.9
Sergeant Colin William Dunning RAAF No. 437405. Age 19	24 Apr 1944 Vc MA353 A58-232	Adelaide River War Cemetery Grave S.B.7
Flying Officer Arthur Keith Kelly RAAF No. 401968 Age 32	18 Sep 1944 VIII JG622 A58-435	Northern Territory Memorial Panel 7
Flying Officer Evelyn Graeme Ivor Etherington RAAF No. 403804. Age 26	20 Nov 1944 Wirraway A20-633	Adelaide River War Cemetery Grave D.C.8
Flying Officer Thomas Claude Smith RAAF No. 409008 Age 25	20 Nov 1944 Wirraway A20-633	Adelaide River War Cemetery Grave D.C.9
Flight Lieutenant John Ridgeway Sturm RAAF No. 407133 Age 26	17 Dec 1944 VIII MT781 A58-478	Townsville War Cemetery Grave B.D.12
Flying Officer Ian Allison Markwell RAAF No. 414054 Age 24	24 Dec 1944 VIII MT620 A58-519	Ambon Memorial Column 8
Leading Aircraftman Ronald Charles Orpen RAAF No. 128679 Age 22	15 Jan 1945 Ground explosion	Ambon War Cemetery Grave 30.C.3
Flight Lieutenant Leslie Allen Leeming RAAF No. 400992 Age 30	18 Jan 1945 VIII MT549 A58-524	Ambon War Cemetery Grave 30.C.4
Flight Sergeant Edmond MacLeod Stevenson RAAF No. 432589 Age 20	14 Apr 1945 VIII JG549 A58-419	Shot down 13 Jan 44 and died of Malaria as PoW Ambon War Cemetery Grave 28.D.2
Flying Officer Clive Everard Miller RAAF No. 420424 Age 23	22 May 1945 VIII MT778 A58-533	Ambon War Cemetery Grave 28.A.15
Flight Lieutenant Alan James Proctor RAAF No. 418174 Age 23	2 Jul 1945 VIII MT545 A58-465	Jakarta War Cemetery Grave 5.F.14

Name	Date	Buried
Flight Lieutenant Norman James Cullen RAAF No. 406070. Age 28	10 Jul 1945 VIII MT618 A58-518	Labuan War Cemetery Grave C.A.3
Leading Aircraftman Kenneth Lewis Gordon RAAF No. 140845. Age 21	18 Jul 1945 Ground explosion	Labuan War Cemetery Grave 27.B.8
Flying Officer Edmund Henry Lee RAAF No. 419722 Age 21	15 Aug 1945 Ground Accident	Labuan War Cemetery Grave 24.D.9
Leading Aircraftman Lewis Thacker Percival RAAF No. 78409 Age 30	26 Feb 1946 Illness after discharge	Brisbane (Lutwyche) Cemetery Anzac Plot 1, Portion 7, Sect 83, Grave 5

Prisoners of War

Name	Date/Aircraft	Details
Sergeant Alexander Cyffin Roberts RAAF No. 402007 Age 24	11 Jul 41 IIa P7562	St. Hippolyte, France, Escaped Returned to UK 8 Oct 41 Later Flight Lieutenant
Flight Lieutenant Justin Hillary O'Byrne RAAF No. 408022 Age 39	9 Aug 41 IIa P7682	Stalag Luft III & Ofl. XXIB POW No. 3735
Warrant officer Archibald Roger Stuart RAAF No. 402141 Age 22	18 Sep 41 Vb P8703	Stalag Luft 383 & VII POW No. 9693
Warrant Officer Augustus Kenneth Try RAAF No. 402264 Age 21	18 Sep 41 Vb W3508	Stalag Luft IV & IIIA POW No. 9700
Warrant Officer Ian Arthur Lace Milne RAAF No. 407078 Age 36	20 Sep 41 Vb AB841	Stalag Luft III, VII & 357 POW No. 9667
Pilot Officer Keith Bruce Chisholm RAAF No. 402150 Age 21	12 Oct 41 Vb W3520	Stalag VIIIB POW No. 202 Escaped 11 Aug 42

SOURCES

RAAF (Office of Air Force History and RAAF Museum)

No. 452 Sqn Miscellaneous Records, June 1943.

Air Board Agenda 1940-46

Air Force Organisation Directive 2010-05. Reorganisation of Number 44 Wing and Formation of Number 452 Squadron and Number 453 Squadron, 8 Nov 2010.

Gooch, Wing Commander J.S., *RAAF Operations: Darwin Area*, Manuscript.

RAAF Aerodrome Guide: Australia and Adjacent Islands, RAAFHQ, Nov 45.

Squadron Operational Record Books (AKA: Form 540. Unit History Sheets) 452, 457 & 54 Sqns.

Short History of US Air Forces in Australia to September 1942.

The National Archives (UK)

Air 50/155/13.

London Gazette Supplement

Imperial War Museum

IWM *Finucane Papers.*

National Archive of Australia

A705, 43/1/586. Anti-G Problems Research Committee.

A705, 231/9/578 Pt 1 Attachment. Establishment – Number 452 Fighter Squadron, 3 July 1941

A705, 231/9/1744 Pt 1 – Establishment of No 80 Wing Headquarters.

A1196, 36/501/200 - Movement Overseas No 452 Squadron

A5954, 524/2. Enemy Air Raid No 54 on Darwin – 2nd May 1943.

A9695, 1018. Joint interview with Squadron Leader K James and Squadron Leader R S McDonald, 23 Aug 1944.

A9845, 274. Spitfire [Capstan] [RAF serial numbers] [Accidents Part 1].

A11095, 100/1/Air. Report on Operation Potshot – 10/10 May 1944.

Australian War Memorial

AWM MSS 1648 – Roberts, Alexander Cyffin Memoir.

AWM PR00346 – Brealey, A. O. (Leading Aircraftman) – Diary transcript.

Books

Bower, Chaz, *Fighter Command 1936-1968: The Story of Britain's crack fighter squadrons*, Sphere Books, London, 1981.

Burton, Doris E., *Wing Commander Brendan (Paddy) Finucane*, ACTS Publication, Melbourne, 10 Feb 1968.

Caldwell, Donald L., *JG26: Top Guns of the Luftwaffe*, Orion Books, New York, 1981.

Clutton-Brock, Oliver, *RAF Evaders*, Grub Street, London, 2009.

Cooper, Anthony, *'Paddy' Finucane and the legend of the Kenly Wing*, Fonthill, London, 2016.

Cooper, Anthony, *Darwin Spitfires: The Real Battle for Australia*, New South Press, Sydney, 2011.

Cooper, Matthew, *The German Air Force 1933-1945: Anatomy of Failure*, Jane's, London, 1981.

Cundy, Ron, *A Gremlin on my Shoulder: The Story of an Australian Fighter Pilot*, AMHP, Loftus, 2001.

Delve, Ken, *The Military Airfields of Britain: East Midlands*, Crowood, Marlborough, 2008.

Delve, Ken, *The Military Airfields of Britain: Northern England*, The Crowood Press, Marlborough, 2006.

Foreman, John, *Fighter Command War Diaries, Vol 2, September 1940 to December 1941*, Air Research Publications, Walton-on-Thames, 1998.

Foreman, John, RAF *Fighter Command Victory Claims, Part 2, 1 January 1941 to 30 June 1943*, Red Kite, Walton-on-Thames, 2005.

Francillon, René, *Japanese Aircraft of the Pacific War*, Putnam, London, 1979.

Franks, Norman L. R., *Royal Air Force Fighter Command Losses of the Second World War 1939-1941*, Midland Publishing, Leicester, 1997.

Gillison, Douglas, *Royal Australian Air Force 1939-1942*, AWM, Canberra, 1962.

Grantham, G, & Bushell, E.J., *Shaft of the Spear: Evolution of the RAAF Technical Services to the end of the War*, self-published, 2004.

Hata, Ikuhiko, Izawa, Yasuho & Shores, Christopher, *Japanese Naval Aces and Fighter Units in World War II*, Airlife, Shrewsbury, 1989.

Herrington, John, *The Air War Against Germany and Italy, 1939-1943*, AWM, Canberra, 1954.

Laird, Malcoln & Mackenzie, Steve, *Spitfire The Anzacs: The RAF Through Colonial Eyes*, Venture Publications, Wellington, 1997.

Lax, Mark, *Malayan Emergency and Confrontation 1950 -1966*, Big Sky Publishing, Newport, 2021.

Lewis, Dr Tom, *The Empire Strikes South*, Avonmore Books, Kent Town, 2017.

Mawer, Granville Allen, *Diary of a Spitfire Pilot over the English Channel and over Darwin*, Rosenberg, Dural, 2011.

Middlebrook, Martin, *The Bomber Command War Diaries*, Midland Publishing, Leister, 1996.

Morgan, Eric B., and Shacklady, Edward, *Spitfire: The History*, Key Books, Stamford, 1987.

Newton, Dennis & Bungey, Richard, *Spitfire Leader: Roert Bungey DFC, Tragic Battle of Britain Hero*, Amberley Publishing, Stroud, 2019.

O'Brien, Graham, *Always There: A History of Air Force Combat Support, APDC, Canberra*, 2009.

Odgers, George, *Royal Australian Air Force 1939-1942*, AWM, Canberra, 1962.

Okumiya, Masatake & Horikoshi, Jiro (with Martin Caidin), *The Zero Fighter*, Cassell, London, 1958.

Powell, Alan, *The Shadow's Edge: Australia's Northern War*, MUP, Carlton, 1992.

Price, Dr Alfred, *The Luftwaffe Data Book*, Greenhill Books, London, 1997.

Price, Dr Alfred, *Spitfire Mark V Aces 1941-45*, Osprey, Oxford, 1997.

Prisoners of War Naval and Air Forces of Great Britain and the Empire 1939-1945, J.B. Hayward and Son, Polstead.

Roberts, Tom, *Wingless: A Biographical Index of Australian Airmen Detained in Wartime*, Self-published, Ballarat, 2011.

Sekigawa, Eiichiro, *Japanese Military Aviation*, Ian Allan Ltd., London, 1974.

Shores, Christopher, *Aces High*, Vol 2, Grub Street, London, 1999.

Shores, Christopher, and Williams, Clive, *Aces High*, Vol 1, Grub Street, London, 1994.

Shores, Christopher, *Those Other Eagles*, Grubb St, London 2004.

Sly, Edward 'Ted', *The Luck of the Draw*, Self-published, 2003.

Stagg, Ross Smith, *Caterpillar Club Survivor: Lost in the Top End, 1943*, TechWrite Solutions, 2007.

Stokes, D., *Paddy Finucane: Fighter Ace*, William Kimber, London, 1983.

Storr, Alan, *RAAF Fatalities in Second World War in RAAF Squadrons and Support Units*, AWM.

Thomas, Andrew, *Spitfire Aces of Burma and the Pacific*, Osprey, Oxford, 2009.

Tramoundanis, Despina, *Units of the Royal Australian Air Force*, Big Sky Publishing, Newport, 2022.

Venn, Walter, *Restore to Service*, AHMP, Loftus, 1999.

Willis, Steve, and Hollis, Barry, *Military Airfields in the British Isles 1939-1945 pt 2 (F-O)*, Enthusiasts Publications Newport Bagnell, 1982.

Wynn, Kenneth G., *Men of the Battle of Britain*, Gliddon Books, Norwich, 1989.

Newspapers, Magazines and Articles

Alford, Bob, 'The 'Dinah' over Darwin', *Flightpath*, Vol 4, No 1.

'Epic Bail Out', *Aviation Heritage*, June 2019.

Hobbins, Peter, 'Engineering the Fight Pilot: Aviators, Anti-G suits and Allied Air Power, 1940-53', *The Journal of Military History*, Vol 84, No. 1, January 2020.

Lax, Mark, 'Goldsmith, Adrian Philip (Tim) (1921–1961)', *Australian Dictionary of Biography*, National Centre of Biography, Australian National University, Canberra, 1996.

Sydney Morning Herald, 5 May 1943.

The (Melbourne) Herald, 16 May 1942, 5 May 1943

Williams, Ross, 'There was movement at the station ..', *Wings*, Spring 2007.

INDEX OF NAMES

Adam, Pilot Officer John 93, 125, 130
Anderson, Sergeant Robert 47
Arnold, General Henry "Hap" 76
Armstrong, Hugo 119
Armstrong, Flying Officer Robert 42, 48, 66, 117
Bader, Wing Commander Douglas 35, 84
Baker, Pilot Officer Al 49, 122, 128
Baldwin, Flight Sergeant CJH 43
Barclay, Squadron Leader Kel 112, 113, 116, 117
Barry, Justice 110
Bassett, Flying Officer Kenneth D 43, 93-95, 125
Batchelor, Lee 69
Beamish, Group Captain Victor 50, 51
Beaton, Pilot Officer Malcolm "Junior" 101, 102, 126
Beattie, Squadron Leader Doug 106
Beauchamp, Monsignor Henry 63
Beech, Sergeant Colin 99
Beeston, Paddy 100
Bevan, Flight Sergeant Ron 43, 52, 53
Birch, Squadron Leader Arthur 104, 105, 112, 116
Bisley, Flight Lieutenant John 71, 82, 94, 117, 119, 124
Bladin, Air Commodore Frank 83
Blumer, Flight Lieutenant 112
Bouchier, Group Captain Cecil 44, 45
Boyd, Wing Commander Bob 50
Brealey, Leading Aircraftman Arthur 75
Brennan, Virgil 119
Brock, Pilot Officer Anthony K 71
Bruce, Stanley 44, 63
Bungey, Squadron Leader Bob 19, 20, 27, 39, 45, 50, 116-118, 122
Bullock, Peter 100
Cable, William 118
Caldwell, Wing Commander Clive 64, 70, 73, 79, 83, 91, 95, 105-107, 110, 112, 116
Chadwick, Flight Sergeant 22
Chapman, Sergeant Geoff 33, 120, 128
Chisholm, Flight Lieutenant Keith 33, 34, 36, 39, 40, 46, 117, 118, 120-122, 131
Chrystal, Pilot Officer Bill 106
Churchill, Winston 55, 56, 61, 62
Clifford, Sergeant Michael 65, 124, 129
Cobby, Air Commodore Harry 110
Coker, Flight Lieutenant Frank 42, 50
Coleman, Major Patrick 45
Colyer 124
Coombes, Pilot Officer Bill 75, 92, 93, 125
Cooper, Flight Sergeant 22
Cormack, Flight Lieuetnant Desmond 117
Costello, Sergeant Andrew 26, 27, 120, 128
Cotton, Professor Frank 90
Cowan, Sergeant Raife 24, 30, 36, 47, 48, 118, 121
Cowell, Flying Officer Gerald 83, 93, 124, 125
Cox, Ken 27
Crompton, Flight Lieutenant Lyn 113
Cross, Sergeant Keith 87, 124, 125
Crossing, Flying Officer Bill 22, 23, 31
Cullen, Flight Lieutenant Norm 112, 126, 131
Cundy, Flight Lieutenant Ron 98, 99, 103, 104, 117, 119
Curtin, John 55, 56, 62, 63
Dippel, *Leutnant* Hans-Georg 51
Donald, Sergeant Jack 30, 36, 47
Douglas, Flight Lieutenant Alfred "Pinky" Graham 20, 22, 27, 36, 39, 118, 121
Douglas, Air Marshal Sholto 44
Downes, Pilot Officer David FK 43, 48, 71
Drakeford, Arthur 56
Duncan, Pilot Officer Colin R 76, 87-89, 125
Dunning, Sergeant Colin 101, 126, 130
Dunstan, Sergeant Bruce 28, 36, 44, 118, 121, 122
Dutton, Squadron Leader Roy 19, 24, 116
Eccleton, Pilot Officer Bill 31, 36, 118, 120, 128
Elliott, Leading Aircraftman Walter 65, 129
Elphick, Pilot Officer John 36, 40-42, 48, 54, 121, 122
Emery, Sergeant John 41, 42, 45, 47, 118, 122, 128
Etherington, Flying Officer "Buster" 104, 130
Evans, Flying Officer Davidson 43, 48, 65, 66, 71, 79, 81, 119, 124
Evatt, Herbert "Doc" 55, 58, 62

Fenegan, Flying Officer John D 71
Finucane, Mrs 63
Finucane, Wing Commander Brendan 4, 5, 20, 21, 24, 25, 27, 30-33, 35-37, 39-42, 44, 47, 50, 62, 63, 74, 117, 118, 120-122
Fiveash, Sergeant Ron "Freddy" 99, 100
Ford, Flying Officer William 55, 73, 123, 124, 129
Foster, Flight Lieutenant Bob 71
Fox, Flying Officer Ken 80, 92, 124
Friend, Flight Lieutenant William 113, 116
Fukuda, Major Takeo 82
Galland, *Oberst* Adolf 33, 44
Gamble, Flying Officer Keith 100, 102, 126
Gazzard, Sergeant Dick 27, 36, 120, 128
Geissmann, Sergeant Bernard 43, 122, 128
Gibbs, Squadron Leader Eric 57, 70, 125
Goldsmith, Flying Officer Adrian "Tim" 71, 74, 80, 81, 93, 117, 119, 124, 125
Goodhew, Sergeant Reg 55, 123, 129
Gordon, Wing Commander JAG 48
Gordon, Leading Aircraftman Ken 112, 131
Gould, Flying Officer John 72
Grant, Jim 23
Gurney, Flight Sergeant Brian 112
Hall, Flying Officer Ed 74, 79, 81, 88, 89, 119, 124, 125
Hamilton, Sergeant Malcolm 51, 123, 128
Hanigan, Sergeant James 27, 37, 121, 128
Harper, Sergeant Frank 51, 122, 128
Haydon, Sergeant Gerry 33, 120, 128
Heaney, Flight Sergeant Arch 82
Hinsley, Cardinal Arthur 63
Hitler, Adolf 17
Hoare, Pat 63
Hoile, Doug 100
Humphrey, Flying Officer Andrew 28, 118, 120
Hutchinson, Sergeant Eric 71, 124, 129
Jackson, Sergeant Edgar 30, 40, 42, 47, 122, 128
James, Flight Sergeant J 48
James, Squadron Leader Ken "Skeeter" 70, 81
Jeffrey, Wing Commander Peter 95
Jerries, Pilot Officer George B 71
Job, Ray 100
Johnson, Fred 100
Jones, Toby 100
Kelly, Flying Officer Keith 103, 104, 126, 130
Kenney, General George 81, 92
King, Flying Officer Jeffrey "Argus" 93, 113, 117, 125, 126
Knowles, Wing Commander Ed 54
Lamerton, Flying Officer Bill 87, 123, 125, 129
Lane-Walters, Flying Officer Denys 34, 47
Lee, Flying Officer Ed 113, 131
Leeming, Flight Lieutenant Les 110, 112, 126, 130
Leigh-Mallory, Air Marshal Trafford 84
Lewis, Flying Officer Donald 27, 33, 36, 42, 49, 120-122, 128
Lloyd, Flying Officer Clive P 48, 74, 79, 90, 124, 125
Lowe, Justice 55
Lumley, Flying Officer Bert 105, 106
MacArthur, General 61, 81, 92, 109
MacDonald, Squadron Leader Ron 75, 81, 93, 97, 100, 116, 117, 119, 124, 125
MacNeil, Pilot Officer Hugh 109
Mahar, Flying Officer Eric 100, 101
Mailey, Sergeant Walt 97
Makin, Flight Lieutenant Paul 27, 51, 90, 94, 122, 125
Manning, Sergeant Charles Frederick 28, 38, 121, 128
Markwell, Flying Officer Ian 106, 107, 126, 130
Marshall, Corporal Charles 41
Matthews, Flying Officer Ern 72
Mawer, Flying Officer Allen 79-81, 93, 119, 124, 125, 130
May, Flight Lieutenant Ken 126
McCann, Fred 27
McIntyre, Pilot Officer 71
McNab, Flying Officer Alex 80, 124, 129
McNamara, Air Vice-Marshal Frank 22, 23, 44, 45, 58, 63
Miller, Flying Officer Clive 112, 126, 130
Milne, Warrant Officer Ian 27, 36, 38, 39, 121, 131
Mölders, *Oberst* Werner 44
Moore, Flight Sergeant Eric 73, 75, 124, 129
Moore-Johnson, Eve 63
Mori, Lieutenant Fumio 72
Morris, Wing Commander Robert 116
Morrison, Sergeant John 43, 48, 54, 123
Murray, Dave 100

Newton, Squadron Leader Rob 57

Nichterlein, Sergeant Willie 80, 83, 124, 129

Norwood, Flight Lieutenant Robin 71

O'Byrne, Pilot Officer Justin 33, 120, 131

O'Connell, Flight Lieutenant Timothey 117

O'Loughlin, Flying Officer Col 99, 100, 102, 126

Orpen, Leading Aircraftman Ron 110, 130

Page, Sir Earle 44, 45

Paget, Squadron Leader John 23

Percival, Leading Aircraftman Lewis 114, 131

Pollard, Pilot Officer John 37

Pretty, Flying Officer Jack 106, 117, 126

Proctor, Flight Lieutenant Alan 112, 126, 130

Reid, Flight Lieutenant Len 114, 116

Richardson, Flight Sergeant Arthur Ross 87, 90, 95, 125

Roberts, Sergeant Alex 26, 27, 46, 120, 131

Roberston, Flight Lieutenant Ken 91, 100, 103

Rommel, Erwin 63

Ross, Flight Lieutenant John 43, 48, 65, 129

Ruskin-Rowe, Pilot Officer Tom 83, 124, 129

Sattler, Flight Lieutenant Geoffrey 101

Schrader, Sergeant Eric 30, 40-43, 47, 122, 128

Schöpfel, Major Gerhard 44

Seldon, Flight Lieutenant William 23, 31, 48, 66, 67

Setoguchi, Lieutenant Kurasuke 72

Sheldon, Pilot Officer ME 48

Shelly, Flight Sergeant FP 48

Sims, Leading Aircraftman Stanley Kirk 93, 130

Sly, Pilot Officer Ray 41, 48, 52, 53, 118, 121, 123

Sly, Flying Officer Ted 53, 54, 111, 112

Smith, Flying Officer Tom 104, 130

Smith, Sergeant WJ 48

Spence, Squadron Leader Lou 97-99, 100, 105, 116, 117

Stagg, Sergeant Ross 71, 80, 124

Stevenson, Bill 100

Stevenson, Flight Sergeant Ed 109, 110, 126, 130

Stockley, Sergeant Henry 71, 124

Strauss, Captain Allison W 69

Stuart, Sergeant Arch 27, 36, 38, 118, 120, 121, 131

Sturm, Flight Lieutenant John 105, 106, 126, 130

Swift, Warrant Officer Clive 117

Tainton, Pat 54, 118, 120

Thorold-Smith, Squadron Leader Ray 4, 5, 24, 27, 33, 37, 40, 41, 43, 44, 49, 50, 54, 70, 71, 74, 75, 90, 116-118, 120-122, 124, 129

Truscott, Squadron Leader Keith 4, 5, 27, 29, 33-36, 39, 40, 42-45, 48, 50, 51, 54, 63, 74, 116-118, 120-123

Try, Sergeant Ken 38, 121, 131

Tuckson, Flying Officer Jerry A 71

Tully, Pilot Officer Paul 73, 79, 93, 125

Turnbull, Flight Sergeant Francis 71

Walters, Group Captain Alan "Wally" 70

Waters, Sergeant Ralph 55, 123, 129

Watkin, Flight Lieutenant Al 93, 129

Wawn, Flight Lieutenant Clive 36, 48, 50, 51, 54, 117, 118, 121, 123

Weatherhead, Flying Officer Ernest 71

Wellsman, Sergeant Jim C 71

Wheeler, Sergeant David M 71

Whillans, Flying Officer Robert 43, 48, 55, 66, 67, 123, 124

White, Pilot Officer Harry 31

Wilkinson, Sergeant WJ 43, 48

Williams, Flying Officer Ken J 114, 116

Williams, Sergeant Ken V 37, 121, 128

Williams, Air Vice-Marshal Richard 44

Williams, Flying Officer Reg R 48, 49, 83, 124

Willis, Pilot Officer William 27, 36, 38, 118, 120, 121, 128

Wise, Leading Aircraftman Maurie 104

Young, Pilot Officer Fred J 71, 92, 125